Early Modern Écologies

Early Modern Écologies

Beyond English Ecocriticism

Edited by
Pauline Goul and Phillip John Usher

Routledge
Taylor & Francis Group
LONDON AND NEW YORK

First published in 2020 by Amsterdam University Press Ltd.

Published 2025 by Routledge
4 Park Square, Milton Park, Abingdon, Oxon OX14 4RN
605 Third Avenue, New York, NY 10158

Routledge is an imprint of the Taylor & Francis Group, an informa business

ISBN: 9789462985971 (hbk)
ISBN: 9781041178460 (pbk)
ISBN: 9781003694397 (ebk)
NUR 617 | 685

Cover illustration: Goldschmied & Chiari, Nympheas #34, 2011, Lambda print 125x333 cm, courtesy of the Artists.

Cover design: Coördesign, Leiden

DOI 10.5117/9789462985971

Table of Contents

Part 3 Groundings

List of Illustrations

Acknowledgments

It is difficult to pin down exactly whom to acknowledge for their help in writing this book since so many diverse conversations led to its existence, but we would like to start by thanking Louisa Mackenzie for setting out, in a single question at the MLA convention, the general line of thought explored in this book. We offer hearty thanks to our contributors for being part of this laboratory of early modern French *écologies*—it is our hope that it will continue to foster productive entanglements between the many nonhuman keywords and quite human persons that interact throughout. We are delighted that this volume appears in the AUP series Environmental Humanities in Pre-modern Cultures, and we thank the series editors (especially Steve Mentz) for their warm welcome. We thank Erika Gaffney for her care and professionalism in ushering this project from its early stages to this final product. Melanie Hackney and Aileen Christensen provided editing and formatting, for which we are grateful. We thank Goldschmied & Chiari (Sara and Eleonora) for granting us the permission to use their beautiful work *Nympheas* for our cover art, and we thank Diane Brown for suggesting that their art may resonate with our project. Pauline Goul would like to thank Karen Pinkus for introducing her to (for lack of a better word) ecocriticism, Kathleen Long for her constant support, and Phillip John Usher for taking on this collaborative effort and generally being an ideal early modern *écological* companion, in more ways than one. Pablo García-Piñar should always be acknowledged, although listing why would require another book. Phillip John Usher expresses his gratitude to Pauline Goul for her immediate excitement about this project and her dedication to seeing it into print. He dedicates his part of the volume to Chloé Juniper Usher, who joined this world as the book came together, giving it a whole new urgency and depth.

Introduction

Pauline Goul and Phillip John Usher

'What can early modern French literature do for ecocriticism?' This was Louisa Mackenzie's question during a roundtable discussion at a recent MLA convention. As she noted, it is a much better and more important question than 'What can ecocriticism do for early modern French literature?' and it caught the attention of the editors of the present volume.[1] Mackenzie's point here is that we should be letting early modern French literature interrogate and shape contemporary theory and criticism, rather than *applying* existing ecocritical paradigms *onto* authors such as Rabelais or Ronsard. After many conversations and several follow-up panels (including one at the Renaissance Society of America's annual conference in Boston, with Mackenzie as chair), this point not only seemed increasingly pertinent, but it had also clearly struck a chord with colleagues who detected a groundswell of interest in re-reading works of early modern French literature from a particular angle. The present volume is the concrete product of this groundswell. Its title (*Early Modern Écologies*) is subtly bilingual, the acute accent (*é*) on the final word drawing attention to the fact that, in method and in conclusions, the chapters that follow are caught between languages and literary and critical traditions. Whether read from an Anglophone or a Francophone point of view, the book as a whole speaks, intentionally, with an accent.

As a whole, the present volume opens up a number of conversations around Mackenzie's compelling question. It is not the first collection of writings about French literature and ecocriticism: it arrives *dans le sillage* of a 2012 *FLS* volume on 'The Environment in French and Francophone Literature and Film' edited by Jeff Persels, a 2015 special issue of *Dix Neuf* titled 'Ecopoetics/L'Écopoétique', edited by Daniel A. Finch-Race and Julien Weber, a 2017 issue of *L'Esprit créateur* titled 'French Ecocriticism/L'écocritique

1 The vanishing point of Mackenzie's MLA talk—and, arguably, of the present volume—is her article, 'It's a Queer Thing: Early Modern French Ecocriticism', which makes a resounding and articulate call for putting early modern French literature into dialogue with questions of ecology.

Goul, P. and P.J. Usher (eds.), *Early Modern Écologies. Beyond English Ecocriticism.*
Taylor & Francis Group, 2020
DOI 10.5117/9789462985971_INTRO

française', also edited by Finch-Race and Weber, and Daniel Finch-Race and Stephanie Posthumus's volume *French Ecocriticism.* It is the first, however, to focus exclusively on the possible connections between *early modern* French literature and contemporary theoretical positions. Within the context of British literatures, of course, scholars have been prolific in asking environmental and ecological questions of early modern literature, as shown by the likes of Bruce Boehrer, Todd A. Borlik, Gabriel Egan, Ken Hiltner, Steve Mentz, Vin Nardizzi, Jeffrey Theis, Robert Watson, Tiffany Worth, and many others.[2] A number of recent conference panels and roundtables have started to bring scholars—including the editors of this volume—into the same room and have brought traditions into dialogue, and we hope that the present volume will generate further conversations and collaborations. While recognizing ourselves in and building on English early modern ecocriticism, we also felt that our own primary texts and the theoretical habits of French departments pulled us in other directions. The noise created by the friction between these different cultures, disciplines, and languages is precisely what we hope to use as we start imagining new cartographies of early modern ecocriticism.

On a theoretical level, one of this volume's key contributions is to show that the 'texture' of contemporary eco-theory *could* have been otherwise—and could still be. Had contemporary theorists such as Timothy Morton and Bruno Latour developed their thought around French-language sources instead of English-language ones, things might have looked a little different. As the authors of the following studies demonstrate, if Timothy Morton had started with Guillaume de Salluste du Bartas or Pierre de Ronsard instead of John Milton, or if Bruno Latour had begun with Jean Bodin or Olivier de Serres instead of Thomas Hobbes and Robert Boyle, then key works and key *words* for ecocriticism and Anthropocene Studies would not be quite the same. This volume, one might say, imagines some shards of these alternative works. More generally, the chapters that follow knowingly enter into a space of reflection that has been dominated for many reasons by the English language and by English-language traditions. The modern

2 Boehrer, *Environmental Degradation in Jacobean Drama*; Borlik, *Ecocriticism and Early Modern English Literature: Green Pastures*; Egan, *Green Shakespeare: From Ecopolitics to Ecocriticism*; Hiltner, *Milton and Ecology*; Hiltner (ed.), *Renaissance Ecology: Imagining Eden in Milton's England*; Hiltner, *What Else is Pastoral? Renaissance Literature and the Environment*; Mentz, *At the Bottom of Shakespeare's Ocean* and *Shipwreck Modernity: Ecologies of Globalization, 1550–1719*; Nardizzi, *Wooden Os: Shakespeare's Theatres and England's Trees*; Theis, *Writing the Forest in Early Modern England: A Sylvan Pastoral Nation*; Watson, *Back to Nature: The Green and the Real in the Late Renaissance*; Hallock, Kamps, and Raber (ed.), *Early Modern Ecostudies: From the Florentine Codex to Shakespeare*; Bruckner and Brayton (ed.), *Ecocritical Shakespeare*.

canon of ecology and ecocriticism is largely Anglophone, as are many of their institutional frameworks. The fact that the MLA's Ecocriticism and Environmental Humanities forum defines itself as 'a scholarly practice *within English Studies*', whatever the degree of intentionality of such a definition, is a symptom of the monolingualism of the Environmental Humanities in the United States.[3]

The collection does not claim to offer a definitive answer about the connection between early modern French literature and ecology—this is why the title is in the plural. Each author asks and explores Mackenzie's original question—'What can early modern French literature do for ecocriticism?'—in relation to texts and specific problems pertinent to their own current research. The volume's scope is wide—but there are a number of voices and topics that return with regularity, and a three-part structure emerged organically over the book's evolution. The path ahead can be summarized as follows. Our authors identified three major theoretical problems that have received much attention. Following this introduction and a 'threshold' article, the book enters its first zone, *Dark(ish) Ecologies*. This section brings together contributions that work through and sometimes challenge ideas related to Timothy Morton's thought, especially his concept of 'dark ecology', in which we humans (as we seek out ecological awareness, get caught up chasing after both ourselves and 'strange strangers': in which we are both detective and criminal. The second section (or laboratory) focuses on the difficulty of negotiating the definitions of—and relationship between—'nature' and 'culture', a huge array of problems most notably formulated by Bruno Latour and Donna Haraway (who speaks of 'naturecultures').[4] The third and final section is an experimental workroom, somewhat in the mood of the *Animer le paysage* exhibition at the *Musée de la chasse et de la nature* in Paris (in 2017) and of Latour's latest *Où atterrir*? This section comprises chapters that focus on ground and grounding.

Ahead of these three parts, however, this collection opens with Hassan Melehy's chapter, 'Off the Human Track: Montaigne, Deleuze, and the Materialization of Philosophy'. Purposely located at the threshold between the editors' 'Introduction' and the ensuing sections, the chapter reviews the contentious history of the relationship of theory with early modern

3 See https://thewire.mla.hcommons.org/ecocriticism-environmental-humanities/. Accessed 19 January 2017. Emphasis added. For a longer discussion of the monolingualism of the Environmental Humanities, see Usher, *Exterranean: Extraction in the Humanist Anthropocene*, 'Introduction'.

4 Haraway, *The Companion Species Manifesto: Dogs, People, and Significant Otherness.*

French literature, including flashpoints such as Tom Conley's 1978 article 'Cataparalysis' in *diacritics*, Terence Cave's *The Cornucopian Text* (1979), and Gérard Defaux's career-long attack on theoretical approaches to the early modern period. Offering more than a history, however, Melehy unpacks the terms of the arguments for or against theory, asserting against a staunch historicist position that '[the] very notion of the reconstruction of a past era that is complete enough to determine a text's meaning is as triumphalist as the caricatures of theory that historicists [...] have routinely made'. As Melehy puts it—and such an assertion clearly undergirds the present collection as a whole—'[in] its best version, theory involves a constant suspicion with regard to its own completeness'. French departments might sometimes be thought of as one of the natural homes of what François Cusset calls *French Theory*, but over the last four or five decades they have also been harboured the outright and caricatured rejection of theory—even in 2018, Edwin Duval could stand in front of a room full of scholars and write off theoretical approaches to French literature as 'forgetting the text' and as dealing 'only with race and gender'.[5] Following on from the Introduction, the first part of Melehy's chapter thus continues the crucial task of situating *Early Modern Écologies* within a specifically French-literature history of criticism, and of pre-empting the criticisms that will likely arise from certain quarters. The second part of 'Off the Human Track' develops and deploys a method for reading early modern French literature and contemporary theory in dialogue, taking up Gilles Deleuze and Michel de Montaigne in particular as two authors who—despite all that separates them, and the former quoting the latter only once—are linked by their anti-Platonism (i.e. the 'rejection of the dominant metaphysics of the West that hierarchizes the relation between thought and reality') and their attempts to re-align perception, thinking, and matter in light of that rejection. The potential of Montaigne's *Essais* as a key text here is that it may be re-inserted into those debates within ecocriticism and new materialism that focus on matter, and which often draw on Deleuze (as Jane Bennett does, for example). Melehy's chapter thus begins this collection with a defence of the volume's theoretical project, a test case of reading in such a way, and a rallying cry for Montaigne's importance for early modern *écologies*.

Turning to Part 1, *Dark(ish) Ecologies* begins with Stephanie Shiflett's study, 'Du Bartas Responding to Morton's Milton: A Bodily Route to the

5 Duval made these comments during a talk at the *Atelier du seizième siècle* held at Tulane University, 21 March 2018. For confirmation of Duval's place in this battle for and against theoretical approaches to the early modern, see Tom Conley, 'Fadaises et dictons', p. 255.

Ecological Thought', which offers a *Frenching* of Timothy Morton's notion of the 'ecological thought' that in turn suggests new coherences and alignments within Morton's lexicon. Shiflett's starting point is Morton's quotation of Raphael's speech to Adam in Book 7 of *Paradise Lost*, a kind of thought journey that articulates and makes possible the 'uber-macrocosmic thinking' central to the 'ecological thought'. Turning to Du Bartas's epic *La Sepmaine*, situated as *Paradise Lost*'s predecessor, Shiflett explores alternative poetic formulations for such thought journeys. She first examines the moment where Du Bartas 'compares God to a painter who steps back to admire his own masterpiece' (after Genesis 2. 2)—whereas Milton's Raphael and Adam stand in the Garden of Eden and look into outer space, Du Bartas's God looks back at Earth from nowhere. This and other passages in Du Bartas produce related moments of scale-shifting that Shiflett argues are essential to the 'ecological thought'. In a final move, Shiflett explores another of Du Bartas's poetic scale-shifting voyages, leading this time into the human body, in another point of connection with Morton: for the ecological thought, 'everything is DNA', whereas for Du Bartas 'everything is [Aristotelian] elements'. Du Bartas's route to the Mortonian 'ecological thought' is not Milton's—but it is precisely their differences that are of interest, as Shiflett re-inscribes Morton's theory into an alternative and French literary history that also recalibrates the theory.

Jennifer Oliver's chapter '"When is a meadow not a meadow?": Dark Ecology and Fields of Conflict in French Renaissance Poetry' opens by quoting one of the most famous of French verses: Agrippa d'Aubigné's statement, 'Je veux peindre la France une mère affligée' ('I want to paint France as a tormented mother'). Fully acknowledging the historical specificity of this and other bodily and visceral images in the poetry of Pierre de Ronsard and Agrippa d'Aubigné—in particular their connection to the political situation of early modern France, and especially to the Wars of Religion that pitted Catholics against Protestants—Oliver studies such images in an attempt to gain 'access to pre-Heideggerian, and indeed pre-Kantian, pre-Cartesian, possibilities for thought'. More specifically, she examines 'uncanny', 'weird', and 'loopy' corporeal poetic images in light of, and with the aim of adding further texture to, the notion of toxic agrilogistic thought that Timothy Morton studies in his *Dark Ecology*.

In her chapter 'Equipment for Living with Hyperobjects: Proverbs in Ronsard's *Franciade*', Kat Addis examines the presence of proverbs in Ronsard's unfinished epic *La Franciade* (1572), which was hugely popular in the early modern period. Addis advances the hypothesis that these short pithy sayings—in their form rather than in their specific content—function ecologically, and that reading them in such a way can train our ear for listening in the Anthropocene. Putting the early modern commonplace tradition into dialogue with Timothy Morton's notion of the hyperobject (deemed viscous, molten, nonlocal, phased, and interobjective), Addis shows how Ronsard's proverbs, which are marked by punctuation in the text since its first publication, force the reader to step away from the *now* to see that moment's connection to hyperobjects such as Fate, and to timelines that outsize the epic. Pausing to listen to proverbs spoken by the sea goddess Leucothea or the prophetess Hyante reveals how they resound as a call to be heard by the *anthropos* of our times as well. More broadly, Addis's contribution to contemporary theoretical debates is an insistence on form over content: to read and to think ecologically is not—and least *not necessarily*—to read or to think *about* something in particular (global warming, environmental degradation, etc.); rather, it is fundamentally a rhetorically governed process.

Pauline Goul's chapter, 'Is Ecology Absurd: Diogenes and the End of Civilization', turns received notions on their head in a manner that recalls a quip from Timothy Morton: 'You think ecologically tuned life means being all efficient and pure. Wrong. It means you can have a disco in every room of your house'.[6] Goul's objective is the rehabilitation of the Greek philosopher Diogenes the Cynic as a thinker of ecology, itself seen as a form of absurdity. To this end, Goul offers a careful reading of a number of Diogenic moments in the writings of François Rabelais and Michel de Montaigne. On the one hand, Goul finds that Rabelais 'portrays the Cynic as a moved and moving man, far from the image of a lazy beggar' and as someone who is very keenly *not* outside of the *polis*, but *outside-within* it. Rabelais's Diogenes is thus seen as engaging in 'an urban ecology of homelessness that is also a humanist cosmopolitanism'. Montaigne's direct treatment of Diogenes, on the other hand, is something of a 'missed encounter'. The true Diogenic moments in the *Essais* are elsewhere: Montaigne 'appears to be most Diogenic when not even bringing up Diogenes'. Bringing these and several other early modern treatments of Diogenes into the critical space of new readings of the absurd (via the work of Carl Lavery and Clare Finburgh), Goul offers a rousing call for dark-ecological absurdist survival that echoes and reshapes the conclusions of Roy Scranton's *Learning to Die in the Anthropocene*.

6 I quote Morton's Twitter comment from Alex Blasdel, '"A reckoning for our species": the philosopher prophet of the Anthropocene', *The Guardian*, June 15, 2017. https://www.theguardian.com/world/2017/jun/15/timothy-morton-anthropocene-philosopher. Accessed 27 September 2019.

Part 2, *Nature's Cultures*, opens with Sara Miglietti's chapter 'Between Nature and Culture: The Integrated Ecology of Renaissance Climate Theories'. Miglietti offers a re-evaluation of what are often called early modern 'climate theories', according to which the human body, mind, and character are shaped by place and climate. Opposing received interpretations that frequently write off such ideas as 'pseudo-science' and geographic determinism, Miglietti turns to the writings of Loys le Roy, Jean Bodin, and Nicolas Abraham de la Framboisière in order to demonstrate how such theories develop a sense of reciprocal relationality between culture and nature—'humans are [...] nature embodied'—in ways that anticipate the ideas of contemporary thinkers such as Philippe Descola. As Miglietti shows, Le Roy's *De la vicissitude ou variété des choses en l'univers* (1575) offers a particularly careful formulation of the interconnectedness of humans and the nonhuman universe, and one which puts emphasis on the non-deterministic character thereof. Bodin's *Methodus* (1566) and *République* (1576) provide material for a more in-depth exploration of the tension between influence (of the nonhuman) and autonomy (of the human), while a reading of La Framboisière's *Gouvernement necessaire à chacun pour vivre longuement en santé* (1600) approaches the nature–culture connection in the context of food and diet. The three early modern French authors studied here call, in Miglietti's reading, for understanding 'climate theory' not as a determinism, but as a form of embeddedness pertinent to our own times.

Phillip John Usher's chapter 'Almost Encountering Ronsard's Roses' takes up the French poet's most famous ode, 'Mignonne, allons voir si la rose...', in order to ask a simple but important question: what are the barriers to close-reading a poem such as this one—a poem made of 'signs'—if we (also) try to access through it the nature or Nature of which it claims to be an imitation? To explore such a question, Usher experiments with three ways of reading the ode. He first explores the cultural/historical approach offered by book history, i.e. by tracing out several steps in the poem's reception in music and in poetic anthologies up to the nineteenth century. A second approach seeks out possible connections between Ronsard's poem and early modern botany's own discussion of roses. A third method, which strives to get beyond the poem as cultural artefact, draws on contemporary plant theory, especially the work of Matthew Hall, Jeffrey Nealon, Michael Marder, and especially Luce Irigaray. Ultimately, Usher strives here to both nudge theoretical discourse away from its zoocentrism and to argue more generally for forms of what Gianni Vattimo calls *pensiero debole*.

Victor Velázquez's 'Renascent Nature in the Ruins: Joachim du Bellay's *Antiquitez de Rome*' looks towards one of the period's most renowned collections of French vernacular verse to place critical emphasis on new sites. The chapter considers the challenges of conservation and the relationship between nature and culture in times when technology, power, and hubris interact. While *Antiquitez de Rome* is traditionally read as 'a text about human-made culture'—a text about Roman civilization falling because of human Romans' hubris—here Velázquez models for us how to be sensitive to its 'surprising meditation on Nature'. In other words, it is not just that Roman buildings have fallen into rubble, and not just that the imagined, remembered, fantasized Rome of antiquity is different from the Rome that Du Bellay encounters while in Italy working as his uncle's secretary. Rather, as the Roman 'palaces lose their shape and meaning' and as they come to 'litter the natural landscapes' the reader must ask whether or not it is possible even to think of anything such as a 'pristine nature before culture'. With the appearance of a 'renewed nature' in the fallen ruins and the 're-emergence of the natural landscape' within the spaces in which capital-*c* Culture fell, categories intermingle. As Velázquez notes at the start of his chapter, such a re-reading serves to remind us of the difference between humanism and anthropocentrism and to refocus our temporal scales.

Part 3, *Groundings*, starts with Oumelbanine Zhiri's 'An Inconvenient Bodin: Latour and the Treasure Seekers'. This chapter seeks and tracks points of contact between the thought of Bruno Latour—especially his *Nous n'avons jamais été modernes* (*We Have Never Been Modern*) and *L'Espoir de Pandore* (*Pandora's Hope*)—and the writings of Jean Bodin. The pairing is well made: Bodin, better and more warmly remembered for his work of political philosophy, the *Six Livres de la République* (1576), than for his *Démonomanie des sorciers* (1580)—which attempts to prove the reality of witchcraft—can appear, in Ann Blair's words, 'Janus-faced', divided between modern and superstitious premodern, exactly the divisions and modernizing claims central to Latour's thought. Zhiri sees in Bodin an exemplar of the premodern who 'inhabits a nature that, far from being the post-bifurcation realm, is entirely worked through by demons, good or evil, allowed to act by God'. The point, however, is not *only* to claim Bodin as a premodern, but rather to allow Bodin to help us explore 'the Latourian opposition between modern and pre-modern views of nature' by foregrounding 'how deeply contested the pre-bifurcation world was itself'. To explore the Latour-Bodin pairing, Zhiri takes up a number of Bodin's different and varied narrative accounts of treasure seeking, looking at how these narratives map the networks made up of, *inter alia*, hunters, treasure, spirits, God, and Satan. As Zhiri concludes, drawing on Jean Céard, the *Démonomanie* thus reveals itself to be a key moment 'in a heated argument' over what N/nature is/was at the very moment of modernity's becoming.

Building on a reading method developed in his *The Graphic Unconscious in Early Modern Writing* (1992) and on his exploration of cartographic literature in *The Self-Made Map* (1996) and *An Errant Eye: Topography and Poetry in Early Modern France* (2011), Tom Conley's chapter 'Reading Olivier de Serres circa 1600: Between Economy and Ecology' draws our attention to one of France's first soil scientists. Conley reads agricultural engineer Serres as 'a thinker of human reshaping of the planet', whose masterpiece, the *Théâtre d'agriculture et mesnage des Champs* (1600), is pertinent in many contexts and has been claimed by various agendas. Though Serres' modern editor Pierre Lieutaghi specifies that the *Théâtre* in no way anticipates 'ecological thought', he ends up also acknowledging how the work 'transmits knowledge that is still valid' and capable of challenging the tenets and theoretical frames of industrial agriculture. Acknowledging the complexity of the work's engagement with 'science, practical endeavor, and aesthetics', Conley offers a close reading of certain key programmatic moments of the work in order to break down the frontiers between page and field, and more generally between word and thing, showing that the *mesnage* of the title (a relative of *management*) is indeed somewhere between economy and ecology. Arguing for the proximity of text and land, Conley's reading might be seen as an early modernist response to Bruno Latour's recent *Où atterrir?* (2017), where we read: 'il faut accepter de définir les terrains de vie comme ce dont un terrestre dépend pour sa survie et en se demandant quels sont les autres terrestres qui se trouvent dans sa dépendance' ('we must accept the need to define life territories as that on which a terrestrial being depends for its survival while asking what/who are the other terrestrial beings who find themselves in that same dependency'). Of the initial task here, Latour says succinctly: 'D'abord décrire' ('First, describe').[7]

Finally, Antónia Szabari's 'Montaigne's Plants in Movement' (whose title plays on Jean Starobinski's canonical *Montaigne en movement*, 1982), offers a careful and nuanced reading of the place of plants in Montaigne's *Essais*. As Szabari notes, much work in Animal Studies has successfully plotted the essayist's questioning of human-animal connections—most recently in Bénédicte Boudou's excellent *Montaigne et les animaux* (2016)—while largely eclipsing the question of plants, a problem signalled and further problematized in Jeffrey Nealon's *Plant Theory*. Asking, 'Can we speak of a botany or botanical thought in *The Essays*?', Szabari maps Montaigne's

7 Latour, *Où atterrir?* p. 120; p. 119. Our translation.

engagement with vegetality—and specifically with plant movement—in the 'Apologie', 'De la cruauté', 'De l'expérience', and elsewhere. As well as detecting Montaigne's interest in plants' 'aliveness', Szabari particularly shows that, whereas 'animals lead Montaigne into an anthropological investigation', plants 'are channels into matter, physics, the observable, the intimate, and the cosmic material world', making the humble plant the 'exemplary figure of animation' in the *Essais*. As such, Montaigne both anticipates the naturalism of the seventeenth-century botanist Guy de la Brosse (who would go on to found the *Jardin des plantes*), and offers an alternative genealogy for the kind of unpredictable movement and 'vibrant matter' central to the theoretical work of Jane Bennett.

Works Cited

Borlik, Todd, *Ecocriticism and Early Modern English Literature: Green Pastures* (New York: Routledge, 2011).

Boehrer, Bruce, *Environmental Degradation in Jacobean Drama* (Cambridge: Cambridge University Press, 2013).

Bruckner, Lynne, and Dan Brayton, eds., *Ecocritical Shakespeare* (Burlington: Ashgate, 2011).

Conley, Tom, 'Fadaises et dictons' in *Montaigne After Theory, Theory After Montaigne*, ed. by Zahi Zalloua (Seattle: University of Washington Press, 2009), pp. 253–263.

Egan, Gabriel, *Green Shakespeare: From Ecopolitics to Ecocriticism* (New York: Routledge, 2006).

Finch-Race, Daniel A., and Julien Weber, eds., *Ecopoetics/L'Écopoétique*, special issue of *Dix Neuf* 19 (2015).

———, eds., *French Ecocriticism/L'écocritique française*, special issue of *L'Esprit créateur* 57.1 (2017).

Hallock, Thomas, Ivo Kamps, and Karen L. Raber, eds., *Early Modern Ecostudies: From the Florentine Codex to Shakespeare* (New York: Palgrave Macmillan, 2008).

Haraway, Donna J., *The Companion Species Manifesto: Dogs, People, and Significant Otherness* (Chicago: Prickly Paradigm Press, 2003).

Hiltner, Ken, *Milton and Ecology* (Cambridge: Cambridge University Press, 2003)

———, ed., *Renaissance Ecology: Imagining Eden in Milton's England* (Pittsburgh: Duquesne University Press, 2008)

———, *What Else is Pastoral? Renaissance Literature and the Environment* (Ithaca: Cornell University Press, 2011).

Latour, Bruno, *Nous n'avons jamais été modernes* (Paris: La Découverte, 1991).

———, *Où atterrir?* (Paris: La Découverte, 2017).

Mackenzie, Louisa, 'It's a Queer Thing: Early Modern French Ecocriticism', *FLS* 39 (2012), 15–42.

Mentz, Steve, *At the Bottom of Shakespeare's Ocean* (London: Continuum, 2009).

———, *Shipwreck Modernity: Ecologies of Globalization, 1550–1719* (Minneapolis: University of Minnesota Press, 2015).

Nardizzi, Vin, *Wooden Os: Shakespeare's Theatres and England's Trees* (Toronto: University of Toronto Press, 2013).

Persels, Jeff, ed., *The Environment in French and Francophone Literature and Film*, special issue of *FLS* 39 (2012).

Theis, Jeffrey, *Writing the Forest in Early Modern England: A Sylvan Pastoral Nation* (Pittsburgh: Duquesne University Press, 2009).

Usher, Phillip John, *Exterranean: Extraction in the Humanist Anthropocene* (New York: Fordham University Press, 2019).

Watson, Robert, *Back to Nature: The Green and the Real in the Late Renaissance* (Philadelphia: University of Pennsylvania Press, 2006).

About the Authors

Pauline Goul is Assistant Professor of French Literature at The George Washington University. She specializes in early modern French literature and environmental criticism, and she has published pieces in the *Forum for Modern Languages Studies* and in volumes like *Global Garbage* and *French Ecocriticism*. She is currently working on a monograph tentatively entitled *The Anxiety of Waste: New World, Environment and Literature in Renaissance France*.

Phillip John Usher Associate Professor of French Literature, Thought and Culture and of Comparative Literature at New York University, where he is also Chair of the Department of French Literature, Thought and Culture. He is the author, translator, or editor of nine volumes. His most recent monograph is *Exterranean: Extraction in the Humanist Anthropocene* (New York: Fordham University Press, 2019).

1. Off the Human Track: Montaigne, Deleuze, and the Materialization of Philosophy

Hassan Melehy

Abstract
Responding to long-standing criticisms that theoretical readings of early modern literary texts are anachronistic, Melehy argues that past and present phenomena may be understood alongside one another, while still respecting both of them. He brings together Gilles Deleuze and Michel de Montaigne through their shared interest in Lucretius. Melehy demonstrates a Lucretian-inflected materialism in Montaigne's *Essais* that implicitly criticizes Platonic conceptions of the primacy of thought over matter, and concomitant human claims to dominate the physical and natural world. Melehy signals intersections between Montaigne's dissident philosophy and Deleuze's materialist re-conception of the history of philosophy in order to point out ways that the essayist's work speaks to questions that are also pertinent to present-day meditations on the environment.

Keywords: Michel de Montaigne, Gilles Deleuze, Lucretius, materialism, literary theory, ecocriticism

One of the challenging stances that this collection adopts—asking what early modern French literature might bring to theory and the present day rather than what theory may bring to early modern literature—implies a critical response to the common conception of theory as 'used' or 'applied'. This has probably been the most widespread model since the early days of what is now called *theory*, which is to say the late 1960s and early 1970s in the US, where the notion of theory was mainly imported into literature programmes despite its origins in European philosophy. The import was often a vulgarized version

Goul, P. and P.J. Usher (eds.), *Early Modern Écologies. Beyond English Ecocriticism.*
Taylor & Francis Group, 2020
DOI 10.5117/9789462985971_CH01

that reduced vast sets of complicated concepts to methodologies that could be put to use for producing even more readings of literary texts. Whereas one of the reasons for the early interest in theory was that it complicated notions of text, writing, reading, and their institutionalization, the 'application' model has always aimed for re-simplification. In this essay, I will respond to the vulgar model by considering the relationship between, on the one hand, recent and contemporary theorization of the phenomenon of matter, and on the other, early modern literature that addresses the same subject.

Theorizing the French Renaissance

Despite its enduring popularity, the application model has long met with explicit contestation from theorists themselves. For example, in a 1978 article on Montaigne and Derrida titled 'Cataparalysis', a pivotal contribution to the history of theory in French Renaissance studies in the Anglophone world, Tom Conley warns that 'the attempt to propose a graft of Montaigne-Derrida would on first sight smack of a recuperative, academic use of deconstruction, but whose cavalier use these pages would like to dispel from the outset'.[1] Conley then explains similarities between the two writers' situations and concomitant relationships to writing, justifying a reading that is more of a rapprochement than a projection of Derridean concepts onto Montaigne. The following year, 'New French theory' or 'NFT', as it was often called, was cemented in the Anglophone world as an approach to French Renaissance literature by Terence Cave's *The Cornucopian Text*, a book whose exceptional research and writing make it a classic of twentieth-century French studies and even literary criticism. Cave shows his astute grasp of the knotty problems of bringing literary theory to Renaissance texts when he writes in his introduction:

> I have not adopted any specific model of analysis—structuralist, Lacanian, Derridian—since to do so would have been to reduce the sixteenth-century texts to the status of local illustrations of a modern theory. Even if, strictly speaking, this predicament can never be wholly avoided, it can be mitigated by the refusal of a single, rigorously determined model.[2]

Besides the fact that any present-day methodology risks imposing itself on early modern texts, the idea is that the texts, rather than the theory,

1 Conley, 'Cataparalysis', p. 42.
2 Cave, *The Cornucopian Text*, p. xvi.

are the central object of interest; if the critic makes efforts to keep theory from becoming a single unified operation, she may then allow herself to be guided primarily by the texts.

These regularly uttered criticisms and caveats notwithstanding, detractors usually present the application model as the whole of theoretical activity, hence bolstering a caricature that enables easy dismissal. In one of the early salvos of what some termed the 'Querelle des Anciens et des Modernes' (the Quarrel of the Ancients and the Moderns)—the opposition between theorists and traditionalists in French Renaissance studies—Gérard Defaux attacked Floyd Gray for dogmatic adherence to an *idée fixe*. In his review of Gray's 1974 *Rabelais et l'écriture*, he accuses Gray of reducing Rabelais's work to a 'pur jeu d'écritures dépourvues de signification' ('a pure play of writing devoid of signification') and, in anticipation of a soon-to-be widespread characterization of all so-called poststructuralist criticism, of emptying the text 'de toute dimension idéologique' ('of any ideological dimension').[3] Defaux made it a career project to level such charges against theorists. In his 1987 *Marot, Rabelais, Montaigne: L'écriture comme présence*, whose very subtitle targets Derrida's critique of logocentrism and its key notion of writing as absence, Defaux takes aim at Cave. Against the argument and evidence in *The Cornucopian Text* that Rabelais's works raise immense problems concerning the relationship of language and meaning, Defaux insists on the basic optimism of Erasmus about the power of language to signify with certainty, an optimism he claims Rabelais shares.[4] From Defaux's perspective, it is as though Cave's motivation had less to do with Renaissance texts than with a simple destructive will. Defaux's position boils down to the idea that those who go along with Derrida and Paul de Man (mainly Cave, but also François Rigolot and Michel Jeanneret) are simply taking a fashionable idea belonging to the late twentieth century, that of the irreducible plurality of meaning, and violently imposing it on sixteenth-century literature.[5]

In presenting itself as a counterpoint to the application model, this collection takes seriously at least some of the problems that its detractors raise. This is appropriate: given that the application model is widely accepted by practitioners of theory, its drawbacks still need to be addressed, even several decades later. It may be that the many ripostes to the application model do not succeed in leaving it behind. It may be that theoretical approaches,

3 Defaux, review of Floyd Gray, *Rabelais et l'écriture*, p. 1050. My translation.

4 Defaux, *L'écriture comme présence*, p. 103. John O'Brien provides a superb account of Cave's relationship to theory in *The Cornucopian Text* and Defaux's criticisms (pp. 14–20).

5 Defaux, *L'écriture comme présence*, pp. 103–104.

by virtue of involving a set of concepts formulated in the recent past (as Cave suggests), always risk anachronism when brought to bear on texts and phenomena from the more distant past. It is easy to see how this might be the case with respect to ecocriticism: awareness of ecology as such and of humanity's responsibility towards it belongs especially to the mid-twentieth century and the decades since, though it certainly has antecedents in more than just romanticism and postromanticism. As Jane Bennett has shown, notions of 'the force of things' and the agencies of matter, which are important in the ecological thinking that brings into serious question human claims to dominion over the earth, have a history in such figures as Spinoza and Lucretius.[6]

Theoretical approaches to literary criticism, which in their early days were marked by declarations of finally getting to the truth,[7] may well draw on a triumphalism similar if not identical to that which informs the science that made large-scale exploitation of the environment possible. The hazards of bringing twentieth- and twenty-first-century theory to Renaissance texts could be considered in some ways comparable to ecological abuse: the approach may run the risk of serious compromise or outright destruction, of imposing an epistemological and technical apparatus onto something it can easily dominate and absorb. On the other hand, treating sixteenth-century texts in their material specificity and as traversed by agencies not temporally or teleologically subordinate to our own is integral to contesting such erasure; it is to respect these texts' engagement with the environment as part of our own critical environment, without insisting that their approaches are simply the ones we are familiar with. The idea of such a project is rather to regard these approaches as a challenge to our own and, in the best of cases, as offering suggestions for pushing against the limits of our own. Treating them as ecological phenomena in themselves may teach greater intellectual sensitivity to ecological phenomena.

But the other side of the problem is the following: how can we in the present know whether what we believe we recognize in Renaissance texts has any relationship to our own concepts? This question merely repeats the

6 See especially Bennett's preface (pp. vii-xix) and first two chapters, 'The Force of Things' (pp. 1–19) and 'The Agency of Assemblages' (pp. 20–38).

7 In his introductory chapter in *Textual Strategies*, a volume that more than most defined theoretically informed criticism, Josué Harari speaks of theory's role in 'the gains and losses, the advancements and retrenchments of criticism' (p. 72). Jeffrey Nealon notes that such phrasing 'seems to replicate rather than displace the violent will-to-truth that is in question in so many of the theoretical discussions [Harari] presents', which include contributions by Barthes, de Man, Derrida, Foucault, Gilles Deleuze, René Girard, and others (p. 106).

objection that historicism consistently makes with regard to theory; the notion of theory is then countered with one version or another of the claim that texts can be best, if not exclusively, understood in connection with their own time. In his 1987 attack on all things poststructuralist, Defaux invokes a text's 'fonction traditionnelle' ('traditional function') of mediating meaning, as though tradition itself were enough to justify the idea that critical attempts to empty texts of meaning, as he polemically puts it, can only be a passing fashion.[8] The main reason, in Defaux's argument, that Rabelais doesn't fundamentally question the capacity of language to convey meaning is his historical proximity to Erasmus's optimistic view of language. This point rests on the notion that texts must be understood primarily in relation to phenomena and circumstances in their time. However, one of the main problems with this idea, as Eric Hayot has signalled, is that 'no reconstruction of the categories [of knowledge] as they were known can proceed, given the difference between the present and the past, without bringing to bear on the past some knowledge that it never had—minimally, for instance, the knowledge of that past's future'.[9] The very notion of the reconstruction of a past era that is complete enough to determine a text's meaning is as triumphalist as the caricatures of theory that historicists such as Defaux have routinely made. In its technicalism, this notion belongs as much to the twentieth and twenty-first centuries as theory, and hence is just as inappropriately 'applied' to early modern texts.

At its best, theory offers not only a set of concepts with which to approach texts and other phenomena, but also a reflection on those concepts and on their ever-shifting relationships to the texts and phenomena that come into its purview. In this best version, theory involves a constant suspicion with regard to its own completeness: an assumption, in fact, that it can never offer complete responses. (This incompleteness is the source of what de Man famously called 'the resistance to theory', a double entendre suggesting that the resistance *against* theory is the displaced symptom of the resistance that belongs *to* theory).[10] In fact, the word 'theory', implying an entity with definite parameters, is less preferable than the words 'theorization' and 'theorizing', which suggest a continuing process.[11] In this sense, to theorize about what early modern texts might offer ecocriticism is to begin by finding places where contact might be made between early modern texts and

8 Defaux, *L'écriture comme présence*, p. 102.

9 Hayot, 'Against Historicist Fundamentalism', p. 1420.

10 De Man, 'The Resistance to Theory', p. 19.

11 Cf. De Man's comments on 'theorization' and language: 'Conclusions', p. 102.

contemporary concepts, allowing for those concepts to change in the course of the contact. Although there is certainly a lot to learn about what texts mean by the circumstances that surround them in their own time, to claim that they are unreadable outside this context is extremely shortsighted: even the strictest historicism has to start with the basic notion that something is comprehensible about early modern texts over several centuries.[12] A principal task of theorizing is to alter and enhance present-day concepts so as to enable more of this 'something' to be available in the present day.

Approaching matter

In order to carry out such theorizing, I will spend the rest of this essay bringing together two writers, one from the sixteenth century and one from the twentieth: Michel de Montaigne and Gilles Deleuze. Both are concerned with how human beings resort to various means to claim dominance over their surroundings and, concomitantly, with how to make portions of those surroundings available outside the image of dominance that human beings create through their social, political, and cultural activities. Although there are major differences between them, I will show in my commentary that there are enough commonalities to justify the rapprochement between them as a way of better understanding both the sixteenth century and the present.

Deleuze spent the first part of his career, from the early 1950s to the mid-1960s, writing a series of author studies on figures in the history of Western philosophy, a practice Michael Hardt has termed 'an apprenticeship in philosophy'. Deleuze repeatedly returned to these authors during the decades in which he elaborated his own philosophical project, beginning with *Différence et répétition* (*Difference and Repetition*, 1968) and *Logique du sens* (*The Logic of Sense*, 1969). The philosophers he treats include Lucretius (in a short essay I will discuss below), Hume, Nietzsche, Kant, Bergson, and Spinoza. In 1988 Deleuze added to the series a book on Leibniz. Deleuze's interest in these figures is what he perceives as their dissident status, their critique or rejection of the dominant metaphysics of the West that hierarchizes the relation between thought and reality.

Deleuze's approach to the history of philosophy stems from traditional continental conceptions of it, whose roots are in Hegel and which are common (albeit with considerable variation) to Nietzsche, Heidegger, the Frankfurt

12 With little irony, Barbara Bowen characterizes Rabelais's books as 'unreadable' outside detailed knowledge of their historical circumstances.

School, and those whom in the Anglophone world we call the postructuralists. From this perspective, the history of philosophy offers a record of how the West thinks, or—to put it somewhat less simply—how the West has depicted itself as thinking. Reading this record reveals a great deal about how cultures, ideologies, states, and societies have taken shape and functioned. This thinking, for want of a better word, informs our institutions and the protocols they impose on our scholarly and pedagogical practices. Reading the history of philosophy in an antimetaphysical or antistatist way is nothing less than a challenge to our contemporary institutions. This important dimension of Deleuze has often been overlooked in the United States, partly owing to the fact that his work has been mainly received in literature departments. Deleuze has tended to be understood, along with the many other so-called poststructuralists, but mainly Derrida and Foucault, as offering a methodology by which critical work may be conducted. Although I am aware of the risk of painting US intellectual activity with too broad a brush, I believe François Cusset (author of the book whose French title is *French Theory*) has a point when he says that Deleuze is often imported in a way that reinstitutionalizes his thinking.[13]

To be fair, a number of studies (including Hardt's *Gilles Deleuze*, Graham Jones and Jon Roffe's 2009 volume *Deleuze's Philosophical Lineage*, and Gregory Flaxman's entry on 'Philosophy' in the second edition of *Gilles Deleuze: Key Concepts*, 2012) demonstrate that this sometimes-shaky characterization of Western philosophy does not pervade Anglophone Deleuze studies. Keeping this integral dimension of Deleuze's work in mind offers a way of further removing him from the application model of theory and hence demonstrating the shortcomings of that model. The name that Deleuze, among others, gives to the set of apparatuses of hierarchization and exclusion that govern the history of metaphysics—apparatuses that in the present eclipse the challenge of these points of dissension in the past—is *Platonism*. The target of his project of 'renverser le platonisme' ('overturning Platonism'), as he puts it in *Différence et répétitition*, is the domination of the Idea or form over matter: he sees this hierarchy as persisting through a distinction of true and false copies of the Idea, true images and simulacra.[14] His critique valorizes the simulacrum as the freeing of thought from its apprehension of things as unified essences. Replacing the notion of *being* that dominates Western philosophical thinking with that of *becoming*, in the opening pages of *Logique du sens* Deleuze writes, 'Le pur devenir, l'illimité, est la matière du simulacre en tant qu'il esquive l'action de l'Idée,

13 Cusset, 'Becoming', passim, but especially p. 360.

14 Deleuze, *Différence*, p. 92.

en tant qu'il conteste à la fois *et* le modèle *et* la copie' ('Pure becoming, the unlimited, is the matter of the simulacrum insofar as it eludes the action of the Idea and insofar as it contests *both* model *and* copy at once').[15] The very structure of his book, a series of 'séries', as he calls them, enacts an argument that departs from linear logic but respects offshoots and abrupt turns; it then serially connects five appendices, each one an author study and each developing a set of points from the main text.

Deleuze mixes texts traditionally classified as philosophical with some of the literary works that, in his judgement, offer ways out of philosophy's impasses: among his essays on Pierre Klossowski, Michel Tournier, and Émile Zola, we find one on Plato and another on Lucretius. In 'Platon et le simulacre' ('Plato and the Simulacrum'), Deleuze argues that even in the dialogic motion of the Platonic text, the distinction of true image and simulacrum breaks down.[16] But his strongest ally is Lucretius, who notably presents his prototypical 'theory of everything' as a long poem, *De rerum natura* ('On the Nature of Things' or 'On the Nature of the Universe'). In 'Lucrèce et le simulacre' ('Lucretius and the Simulacrum'), Deleuze opposes the Lucretian simulacrum to the Platonic one; the former is not a false image but rather the only source available with which to think about what exists in the world of things. In Book 4 of *De rerum natura*, Lucretius explains how perception works—a process that necessitates the simulacrum. The latter is 'a sort of outer skin perpetually peeled off the surface of objects and flying about this way and that through the air', making contact between objects and thought possible.[17] And Deleuze emphasizes that a further kind of simulacrum is at work in perception: the 'phantasme', the kind that make giants' faces or mountains appear in the clouds. According to Deleuze, these simulacra 'jouissent d'une haute indépendance à l'égard des objets et d'une extrême mobilité, d'une extrême inconstance dans les images qu'ils forment (parce qu'ils ne sont pas renouvelés par des apports constants émis par l'objet)' ('enjoy a high degree of independence with respect to objects and an extreme mobility, or an extreme inconstancy in the images which they form (since they are not renewed by the constant supplies emitted by the object)').[18] This type of simulacrum makes impression and sensation part of thinking, hence undermining any idea that thinking is an apprehension of the true essence of things. Simulacra are integral to the movements of

15 Deleuze, *Logique*, p. 10; Deleuze, *Logic*, p. 2.

16 Deleuze, 'Platon', pp. 300–307; see also Melehy, 'Images Without'.

17 Lucretius, p. 95.

18 Deleuze, 'Lucrèce et le simulacre', p. 319; Deleuze, 'Lucretius and the Simulacrum', p. 275. Deleuze's reference is to Book 4 of *De rerum natura*: Lucretius, p. 98.

atoms that make up all matter: by endlessly combining and recombining, atoms defy the very notion of a unifying essence. Hence, perception and thinking take place on a continuum of undulating, varying matter that entails no neat distinction between the mind and things: Deleuze presents Lucretian materialism as an exemplary anti-Platonism.

Montaigne and the simulacrum

Although (to my knowledge) Deleuze mentions Montaigne in only one plac—a parenthesis in *Logique du sens* that reflects a classical twentieth-century education in the history of philosophy—their shared interest in Lucretius alone suggests that reading them together may yield something of value.[19] The reading of Deleuze alongside Montaigne that I am proposing starts from this perspective. Such a reading might better be described as proceeding to Montaigne by way of Lucretius, to the latter of whom Deleuze offers access. It is also, then, an anti-Platonist reading that connects with the distinct anti-Platonic metaphysics in Montaigne's text, which I will also discuss. The 1989 discovery of Montaigne's heavily annotated copy of *De rerum natura*, published in a superb edition by M.A. Screech in 1998, shows the immense degree to which Montaigne was immersed in Lucretius. The copy is the smoking gun for what Pierre Villey recognized at the turn of the twentieth century: that the *Essais* bear not only a 150 or so quotations from Lucretius but also many Lucretian-inflected accounts of matter and thinking. Montaigne's heavy annotation on simulacra in Book 4 of *De rerum natura* shows the seriousness with which he took this purely material account of the interaction of the mind and matter, and suggests his interest in, if not his acceptance of, an anti-Platonist metaphysics and epistemology.[20] When Montaigne uses the word *forme*, often enough it is hard to escape an aside to Platonic metaphysics, which was widely known in

19 Deleuze, *Logique du sens*, pp. 163–64: 'il faudrait évaluer par exemple la différence entre les *Essais* de Montaigne, qui s'inscrivent déjà dans le monde classique en tant qu'ils explorent les figures les plus diverses de l'individuation, et les *Confessions* de Rousseau, qui annoncent le romantisme en tant qu'elles sont la première manifestation d'une personne ou d'un Je'. Deleuze, *The Logic of Sense*, p. 138: 'it would be necessary to evaluate, for example, the difference between Montaigne's *Essays*, already inscribed in the classical world insofar as they explore the most diverse figures of individuation, and Rousseau's *Confessions*, announcing Romanticism insofar as they constitute the first manifestation of a person, or an *I*.' (In the English translation, the phrase is not between parentheses).
20 Screech, *Montaigne's Annotated Copy of Lucretius*, pp. 137–158.

the French Renaissance through Ficino and turns up as a source of playful nods in works such as Maurice Scève's *Délie* (the title being an anagram of 'l'Idée', 'the Idea'). *Forme* and *idée*, rendered in English as 'form' and 'idea', both translate the Greek word *eidos* in the Platonic texts. Montaigne prefers *forme* when metaphysics enters his discussions, which it does in often oblique ways; in the case I am about to discuss, this choice allows him to simultaneously respond to Aristotelian metaphysics, insofar as the latter's treatment of matter remains in the Platonic paradigm of the priority of form over matter.

According to Villey's dating of the composition of the *Essais*, 'De l'oisiveté' ('Of Idleness'), Chapter 8 of Book 1, is among the first that Montaigne wrote when he retreated to his tower in 1571.[21] Readers are familiar with it as Montaigne's description of his writing process: his programmatic statement, his definition of the essayistic itinerary that he will pursue over many pages.[22] In this chapter the word *forme* comes up negatively in the first sentence as the word 'informes':

> Comme nous voyons des terres oisives, si elles sont grasses et fertilles, foisonner en cent mille sortes d'herbes sauvages et inutiles, et que, pour les tenir en office, il les faut assubjectir et employer à certaines semences, pour nostre service; et comme nous voyons que les femmes produisent bien toutes seules, des amas et pieces de chair *informes*, mais que pour faire une generation bonne et naturelle, il les faut embesoigner d'une autre semence: ainsin est-il des espris.

> (Just as we see that fallow land, if rich and fertile, teems with a hundred thousand kinds of wild and useless weeds, and that to set it to work we must subject it and sow it with certain seeds for our service; and as we see that women, all alone, produce mere shapeless masses and lumps of flesh, but that to create a good and natural offspring they must be made fertile with a different kind of seed; so it is with minds).[23]

21 Villey, in Montaigne, *Essais*, vol. 1, p. 32.

22 I have not seen a critical analysis of 'De l'oisiveté' that treats it as bearing on metaphysics and epistemology. Notable recent studies include those of Krause and Worth-Stylianou: Krause treats *oisiveté* from the perspective of moral philosophy and cultural practice, while Worth-Stylianou views it as a problem in the intersection of literature and medicine. Some of the latter's remarks on Montaigne's gender destabilization in his rewriting of Plutarch (pp. 298–299) are close to mine. From the point of view of rhetoric and the production of writing, both Mathieu-Castellani, pp. 26–43, and Regosin, pp. 154–163, are of value.

23 Montaigne, *Essais* (1999), 1.8.32, my emphasis; Montaigne, *The Complete Essays*, 1.8.20–21.

In a pair of rhetorically bumpy similes, the fantastical idea of women producing unformed flesh is adapted from another of Montaigne's main ancient sources: Plutarch, who relies on an account of generation in which the male imprints the form on the matter that the female provides.[24]

Through the womb-field metaphor, which runs rampant in antiquity and to which we owe the words *sperm* and *semen*, Montaigne slyly takes on both Aristotelian and Platonic accounts of generation. My claim is not that Montaigne's text directly engages certain passages from his ancient predecessors, but rather that his account of idleness interacts with certain philosophical concepts whose prominence in Plato and Aristotle suggests a major role in the history of Western thought. In Book 8 of Plato's *Laws*, the Athenian speaks of the 'law of restricting procreative intercourse to its natural function by abstention from congress with our own sex, with its deliberate murder of the race and its wasting of the seed of life on a stony and rocky soil, where it will never take root and bear its natural fruit, and equal abstention from any female field whence you would desire no harvest'.[25] The 'natural fruit' of intercourse will be, as for Montaigne, 'bonne et naturelle' ('good and natural') generation, the law of society in accord with that of nature. Both male-male intercourse and nonprocreative male-female intercourse are breaches of the natural law of generation and the law that will ideally perpetuate society. In a much-discussed section of *Generation of Animals*, Aristotle extends the metaphor until it no longer is one, since womb and field are both manifestations of the same metaphysical principle. Semen carries the form to the matter produced by the womb: man, then, offers the soul and woman the body of the resulting child; similarly, 'it is the soil that gives to the seeds the material and body of the plant'.[26] In the *Economics*, Aristotle ranks agriculture as the highest art (*tekhnē*).[27] And in the *Nichomachean Ethics*, he characterizes art as that by which the maker brings purposeful form to the thing made: 'art is [...] concerned with contriving and considering how something may come into being which is capable of either being or not being'.[28] In short, if form and matter do not come together through masculine *tekhnē*, there will be no yield. In the case of both Plato and Aristotle, the field and the womb remain barren and unproductive if masculine agency does not intervene.

24 Mathieu-Castellani, *Montaigne*, p. 27.

25 Plato, *Laws*, 838e–839a.

26 Aristotle, *Generation of Animals*, Book 2, Chapter 4, 738b.

27 Aristotle, *Economics*, Book 1, Chapter 2, 1343a–b.

28 Aristotle, *Nichomachean Ethics*, Book 6, Chapter 4, 1140a.

In contrast, for Montaigne, 'terres oisives' ('fallow lands') are fertile and productive even if their yields are 'sauvages et inutiles' ('wild and useless'). By the same token, women supposedly produce, without the aid of semen, 'des amas et pieces de chair informes' ('shapeless masses and lumps of flesh')—such women aren't strictly speaking barren, although their offspring lack the form that paternal art would confer. In light of this statement's slim basis in medical tradition, Maurice Rat notes Montaigne's source as Plutarch's *Conjugalia Praecepta* (*Advice to the Bride and Groom*):[29] Montaigne knew this book in Estienne de La Boétie's translation, the 1571 publication of which he arranged (and for which he wrote the dedications and note to the reader), as well as in Jacques Amyot's 1572 translation. Montaigne's word choices suggest a closer affinity to Amyot's version, according to which women 'sans avoir la compagnie de l'homme' ('without male company') produce 'des amas sans forme de creature raisonnable, ressemblans à une piece de chair' ('masses without the form of a rational creature, resembling a piece of flesh'):[30] unseeded female matter may grow without direction. But Montaigne embellishes the phenomenon Plutarch describes, adding details to what otherwise amounts to a small point made in passing. In Plutarch's text it mainly functions as a simile, bolstering the claim that 'il faut bien avoir l'œil à ce, que le mesme n'adviene en l'ame et en l'entendement des femmes. Car si elles ne reçoivent d'ailleurs les semences de bons propos, et que leurs maris ne leur facent part de quelque saine doctrine, elles seules à par engendrent et enfantent plusieurs conseils extravagantes' ('one must be cautious that the same thing does not happen in the soul and mind of women. For if they do not receive the seeds of good ideas, and their husbands do not share healthy doctrine with them, left all alone they engender and give birth to a number of extravagant notions').[31] Montaigne borrows Plutarch's simile but alters it with his own seed, so to speak, presenting it as a matter of course that women's bodies would be productive without assistance from male agency.

Generations of antiquity

Montaigne's statement gently challenges Aristotelian tradition, in which the male embodies the active, formative principle and the female the passive,

29 Montaigne, *Œuvres complètes*, p. 1437, 33n4.

30 Plutarch, *Œuvres*, p. 380. My translation.

31 Plutarch, *Œuvres*, p. 380. My translation. La Boétie renders the passage similarly, but with fewer words that match Montaigne's text: Plutarch by La Boétie, p. 88.

receptive one. In *Generation of Animals* Aristotle characterizes both semen and menstrual fluid as residues (distinct from waste-products): semen is the final stage of residue in its formation from the bodily nutriments, and hence is the most useful and effective of the residues.[32] Menstrual fluid is also a useful residue; it is, however, a step below semen in its effectiveness and requires semen in order to participate in production. 'It is through a certain incapacity that the female is female', says Aristotle, 'being incapable of concocting the nutriment in its last stage into semen'.[33] In the generation of an embryo, sperm provides the form (*eidos*) and menstrual fluid the matter: 'the menstrual blood is semen not in a pure state but in need of working up, just as in the formation of fruits the nutriment is present, when it is not yet sifted thoroughly, but needs working up to purify it'. Menstrual discharge, a liquid and not solid flesh, is unformed human matter. Without the intervention of the masculine principle, a woman cannot, strictly speaking, be productive—semen is in actuality and menstrual fluid is in potentiality.[34]

But Montaigne characterizes women as productive without the male semen or seed, not hesitating to term their production 'informes' ('shapeless' in Frame's translation, but more accurately 'formless'). In the same fashion, 'terres oisives' ('fallow fields') can be highly productive without the aid of formative human (masculine) *tekhnē*. These characterizations are out of step with Aristotelian metaphysics, which requires that there must be a form working on matter for production to take place. They are also contrary to the law that Plato lays down in the *Laws*, which in accord with nature assures a participation in the form transmitted by masculine agency. Montaigne finds extensive productivity in places where it would be disallowed by a dominant strain of the Western metaphysical tradition—in which form is prior to matter, masculinity to femininity, actuality to potentiality, activity to passivity, movement to rest. As excess and excrescence, such growth is in effect monstrous, a case of impropriety. In other words, idle women and lands behave as though they were not the property of human males. Hence, in the areas that dominant Western theory and practice place under the domination of *tekhnē*, Montaigne views alternative productivities at work.

32 Aristotle, *Generation of Animals*, 725a-726a.

33 Aristotle, *Generation of Animals*, 728a.

34 Cf. Aristotle, *Generation of Animals* 729b: 'The female, as female, is passive, and the male, as male, is active, and the principle of the movement comes from him. Therefore, if we take the highest genera under which they each fall, the one being active and motive and the other passive and moved, that one thing which is produced comes from them only *in the sense in which a bed comes into being from the carpenter and the wood, or in which a ball comes into being from the wax and the form*' (my emphasis).

The impropriety of the dual simile is further compounded when Montaigne finally arrives at the object of comparison:

> ainsin est-il des espris. Si on ne les occupe à certain sujet, qui les bride et contreigne, ils se jettent desreiglez, par-cy par-là, dans le vague champ des imaginations.
>
> (so it is with minds. Unless you keep them busy with some definite subject that will bridle and control them, they throw themselves in disorder hither and yon in the vague field of imagination).[35]

Now, in Plato, it is in the soul that apprehension of the form (*eidos*) takes place, and through such apprehension true knowledge occurs.[36] Accordingly, men who will be the guardians of the law, that by which human society is maintained in accord with nature, 'will need a real knowledge [that is, knowledge involving the forms] of [...] all [matters of import]; they must be able to expound this knowledge in their speech and to conform to it in their practice'.[37] Although Aristotle is highly critical of Platonic metaphysics as assigning an independent existence to the forms (a formulation that Platonists have long disputed), he continues to assign priority to form over matter.[38] Aristotle's epistemology hence develops Plato's in

35 Montaigne, *Essais* (1999), 1.8.32; Montaigne, *The Complete Essays*, 1.8.21.

36 Plato, *Phaedo* 65d–66a. Cf. Irwin, p. 151, and Dancy, pp. 245–247.

37 Plato, *Laws* 966b.

38 Cf. Ross, pp. 157–158: 'The main point of [Aristotle's polemic against the Platonic forms] is this: – The world which is given to us in experience is a world of concrete individual things acting and reacting on each other. In contemplating these we become aware of characters common to many individuals. These are for Aristotle as real, as objective, as the individuals. They are not in any sense the work of the mind, any more than are the Forms to Plato. But he warns us to assign to them a mode of existence which is proper to universals, viz. existence as characteristic of individuals. We must not posit a separate world of universals. And we must not suppose that we can explain the world, which is a world of change, by the operation of mere universals [...]. It may be doubted whether Plato thus "separated" the universal from its particulars. To distinguish the universal from its particulars is in a sense to separate it. It is to think of it as a *distinct* entity. Whether Plato also thought of it as a *separately existing* entity, it is hard to say.' And in a more recent treatment of the Platonic dialogues: 'It should especially be noted that in this statement [*Phaedrus* 249b-c] human understanding is described as a *gathering*, as a *collecting of many into one*. The many that are gathered up in such a gathering are explicitly identified as perceptions [...], presumably in the sense of something made present to the senses [...]. In gathering up any such many into a one, it is necessary that the one into which the many are to be gathered be itself directive in the gathering. In other words, it is necessary to gather *according to* the one of the gathering. Socrates' statement gives a name to such gathering into one according to the one. It is named: according to the *eidos*. This means that an *eidos* is, in the first instance, the one of

that he understands the soul as not only 'capable of receiving the form of the object' and capable of giving form to objects by way of *tekhnē*, but also constituting the form with respect to the matter of the body.[39] The soul, the form, is the masculine principle of movement and actuality, matter and body the feminine principle of rest and potentiality. In his extravagant simile Montaigne places the mind on the side of matter, bodies, unworked fields, and women, and attributes a movement and productivity to all of them. Furthermore, in adapting Plutarch's simile, he attributes to male minds purported characteristics of female minds. That is, in Montaigne's treatment the mind is something other than what it has been in dominant currents of the history of Western philosophy, and is moreover capable of moving and apprehending phenomena outside of Western *tekhnē*.

If the mind is now formless matter, or closely akin to it, it dodges Platonic metaphysics by being itself incapable of apprehending forms, and hence of knowing the truth. However, since Montaigne is willing to admit positive, knowable existence to formless weeds, pieces of flesh, and the mental images he compares them with, it becomes apparent that in his view things do not need to participate in the form or idea in order to be real and functional. That is, something that in Platonic metaphysics would simply be a false image or *phantasma* or simulacrum because it does not participate in the form, for Montaigne is available to the expanded notion of knowledge he is proposing, for the mind that itself is without full form. This mind, proceeding in disorderly directions, never apprehends the whole of anything but is able to attribute existence and partial knowability to all that it encounters. A few lines later, Montaigne finishes the short essay—which is less than a page long in the 1595 posthumous edition—with a description of the operations of the mind in relation to the things that it apprehends in the world:[40]

> Dernierement que je me retiray chez moy, deliberé autant que je pourroy, ne me mesler d'autre chose que de passer en repos, et à part, ce peu qui

a gathering, a one into which and according to which many things made present to the senses are gathered. This connection between *eidos* and the gathering of many into one is of utmost importance as point of departure for thinking through the sense of *eidos* in the Platonic dialogues' (Sallis, pp. 149–150). See also Ricœur, p. 10: 'Essence wins out over things so as to gather them together, to collect them. Here the ontological intention does not yet separate essence from things: in the early dialogues, the relation of inherence is employed but not reflected upon; this is why the Platonic vocabulary remains hesitant: essence is located within, "en," or throughout, *dia*. Identiy is present "in" variation, unity circulates *throughout* the cases (*Meno* 74a, 77a: Plato is led to say there that essence is the *all* of multiple things, which will become the Aristotelian "universal").'

39 Aristotle, *On the Soul*, 429a.

40 Montaigne, *Essais* (1595), 1.8.16–17.

> me reste de la vie: il me semblait ne pouvoir faire plus grande faveur à mon esprit, que de le laisser en pleine oysiveté, s'entretenir soy mesmes, et s'arrester et rasseoir en soy: ce que j'esperois qu'il peut meshuy faire plus aisément, devenu avec le temps plus poisant, et plus meur. Mais je trouve,
>
> *variam semper dant otia mentem* [l'oisiveté dissipe toujours l'esprit en tous sens],
>
> que au rebours, faisant le cheval eschappé, il se donne cent fois plus d'affaire à soy mesmes, qu'il n'en prenoit pour autruy; et m'enfante tant de chimeres et monstres fantasques les uns sur les autres, sans ordre et sans propos, que pour en contempler à mon aise l'ineptie et l'estrangeté, j'ay commancé de les mettre en rolle, esperant avec le temps luy en faire honte à luy mesmes.
>
> (Lately when I retired to my home, determined so far as possible to bother about nothing except spending the little life I have left in rest and seclusion, it seemed to me I could do my mind no greater favor than to let it entertain itself in full idleness and stay and settle in itself, which I hoped it might do more easily now, having become weightier and riper with time. But I find—
>
> Ever idle hours breed wandering thoughts
> LUCAN
>
> —that, on the contrary, like a runaway horse, it gives itself a hundred times more trouble than it took for others, and gives birth to so many chimeras and fantastical monsters, one after another, without order or purpose, that in order to contemplate their ineptitude and strangeness at my pleasure, I have begun to put them in writing, hoping in time to make my mind ashamed of itself.)[41]

One thing Montaigne does in these lines is to underscore the mixture of metaphors that runs through this chapter. The idle land and the idle womb come together in their metaphorical transposition to the idle mind, idleness becoming a condition of uncontrolled, and perhaps uncontrollable, movement and immense productivity. Montaigne's mind has become 'avec le temps plus poisant, et plus meur' ('weightier and riper with time'), like the fruit of

41 Montaigne, *Essais* (1999), 1.8.32; Montaigne, *The Complete Essays*, 1.8.21.

a wild tree. But it does not simply fall into place and allow easy harvesting; instead it becomes a runaway horse, not fixed by the bridle that Montaigne metaphorically slips into his description of minds in the opening of the essay. And in its motion it gives birth to monstrous children. This mind, then, is in constant motion, in keeping with Platonic and Aristotelian conception—and it is productive, but in a way that borrows from femininity. According to classical notions, however, a woman is passive, not in motion, and her giving birth is not strictly her own productivity but also that of the man who has impregnated her. Montaigne's mind, or the exemplary mind he presents here, is a hybrid of masculine and feminine principles, itself fantastic and monstrous as it gives birth to the fantastic and monstrous children that the *Essais* record.

This is the activity and the passivity of an idle mind, an *esprit oisif. Oisiveté* itself, as Montaigne describes it, is a perhaps monstrous hybrid of activity and passivity. The idle mind is an unbridled horse dashing across the idle field, so it is not idle in the sense of remaining at rest. But the purpose of this motion is, ultimately, the idleness of rest. In order to counter the unbridled motion of the idle mind, Montaigne sets to work at 'put[ting his thoughts] in writing', 'les mettre en rolle', an expression that means both to put on paper and to set in order. But this order is not an order, since it preserves the very disorder that might make the mind ashamed of itself. The idle mind is really a mind in motion that produces these chimeras and fantastical monsters in its constant interaction with things, things and mind being both active and passive: these images do not stem from the form of a thing but rather fly from it in the motion Montaigne describes, and hence are akin to the simulacra of Lucretius, related by contiguity to things and then, through motion, becoming contiguous with the mind. Along the lines of Lucretius, Montaigne undermines the distinction between imagining and thinking, rendering both of them a result of the constant motion of matter and the endless combinations and recombinations of portions of matter. As such, the Montaignian mind doesn't dominate matter, but rather continually interacts with it as a part of it. Montaigne's writing is an imitation of this process that is also a participation *in* it: it presents an image of the process that makes it more available to knowledge, and itself becomes a constant process of recombination of matter. In one of his more celebrated descriptions of the *Essais*, from the beginning of 'De l'amitié' ('Of Friendship'), Chapter 28 of Book 1, Montaigne presents his writing as such a process:

> Que sont-ce icy aussi, à la verité, que crotesques et corps monstrueux, rappiecez de divers membres, sans certaine figure, n'ayant ordre, suite, ny proportion que fortuite?

> (And what are these things of mine, in truth, but grotesques and monstrous bodies, pieced together of divers members, without definite shape, having no order, sequence, or proportion other than accidental?)[42]

His writings are an image of his thoughts, both of them an image of the constant motion of matter in the world, and of the recognition on the part of the mind that the privileged place it tends to accord itself is no less of an illusion than the other illusions it encounters. In Montaigne's usage, following Lucretius, images and thoughts themselves behave materially in their interactions with things. Montaigne recognizes that things have a power comparable to that of minds—Bennett, reading Lucretius, Deleuze, and a few others, terms this 'thing-power'.[43] He also accepts that minds may in turn find their proper place among things.

Tracking humans

When Montaigne states his hope that the written images of his mind's own chimeras and monsters will eventually make it ashamed of itself, he jokingly suggests a humbling process by which his mind may try to rein itself in. But since he states this as a wish, it may not happen, and given his continued work on the *Essais* for twenty years after writing these words, it does not look as though the hope is ever fulfilled. At the same time, through the humbling process, Montaigne's mind may become more aware of its small place in the universe, of its role as just one set of component parts of a vastly larger set of operations. His statement on the purpose of education in 'De l'institution des enfans' ('Of the Education of Children'), Chapter 26 of Book 1, suggests this sort of humbling, again as an image that is also a text, in which the mind may see itself and its own status as a tiny component:

> qui se presente, comme dans un tableau, cette grande image de nostre mere nature en son entiere magesté; qui lit en son visage une si generale et constante varieté; qui se remarque là dedans, et non soy, mais tout un royaume, comme un traict d'une poincte très delicate: celuy-là seul estime les choses selon leur juste grandeur.

42 Montaigne, *Essais* (1999), 1.28.183; Montaigne, *The Complete Essays*, 1.28.135.

43 Bennett, *Vibrant Matter*, pp. 20–21.

(whoever considers as in a painting the great picture of our mother Nature in her full majesty ; whoever reads such universal and constant variety in her face ; whoever finds himself there, and not merely himself, but a whole kingdom, as a dot made with a very fine brush ; that man alone estimates things according to their true proportions).[44]

The *Essais* are the book that allows this kind of perspective, part of which is a view of the entities other than human minds that act on the world, work with it, and make sense of it. In the 'Apologie de Raimond Sebond' ('Apology for Raymond Sebond'), Chapter 12 of Book 2, Montaigne famously attributes thinking abilities to animals, in passages that were to trouble Descartes for their questioning of human domination of the world.[45] This section in the longest chapter of the *Essais* begins as a follow-up to the identification of 'la presomption' ('presumption') as 'nostre maladie naturelle et originale' ('our natural and original malady') with Montaigne's remark about his cat: 'Quand je me jouë à ma chatte, qui sçait si elle passe son temps de moy plus que je ne fay d'elle?' ('When I play with my cat, who knows if I am not a pastime to her more than she is to me?').[46] Derrida, placing the 'Apologie' in the history of philosophy as 'l'un des plus grands textes précartésiens ou anticartésiens qui soient sur l'animal' ('one of the greatest pre- or anti-Cartesian texts on the animal that exists'), notes that, in the many examples Montaigne provides of animals who show apparent signs of reasoned action, the question is less that of the animals themselves than it is of the limits of human reason.[47] Nonetheless, it is exactly in recognizing these limits that it becomes possible to imagine other types of reasoning entirely, ones which human beings as yet have little to no ability to reach with their own reasoning. Hence, the world may well be acted on by animals and other kinds of sentient agencies.

In the opening pages of 'Des cannibales' ('Of Cannibals'), Chapter 31 of Book 1, Montaigne presents one of these as a metaphor—though like his suggestions about animals, it is worth taking literally and seriously. Following a discussion of what might have happened to the mythical land of Atlantis and what its relationship might be to the so-called 'New World', in the 1588 B layer of the *Essais* he adds some observations about shifts of land. There are movements in bodies of land, he says, 'en ces grands corps

44 Montaigne, *Essais* (1999), 1.26.157; Montaigne, *The Complete Essays*, 1.26.116.

45 Descartes, pp. 302–304; Melehy, 'Silencing'.

46 Montaigne, *Essais* (1999), 2.12.452 ; Montaigne, *The Complete Essays*, 2.12.330–331.

47 Derrida, 'L'animal', p. 256; Derrida, 'The Animal', p. 6.

comme aux nostres' ('in these great bodies, just as in our own'), beginning a rapprochement between land and human bodies.[48] He writes of having seen the River Dordogne erode the land and buildings along its right bank, and he also reports a similar phenomenon involving severe damage done by the sea:

> En Medoc, le long de la mer, mon frere, Sieur d'Arsac, voit une sienne terre ensevelie soubs les sables que la mer vomit devant elle; le feste d'aucuns bastimens paroist encore; ses rentes et domaines se sont eschangez en pacquages bien maigres. Les habitans disent que, depuis quelque temps, la mer se pousse si fort vers eux qu'ils ont perdu quatre lieuës de terre. Ces sables sont ses fourriers: et voyons des grandes montjoies d'arène mouvante qui marchent d'une demi lieue devant elle, et gaignent païs.
>
> (In Médoc, along the seashore, my brother, the sieur d'Arsac, can see an estate of his buried under the sands that the sea spews forth; the tops of some buildings are still visible; his farms and domains have changed into very thin pasturage. The inhabitants say that for some time the sea has been pushing toward them so hard that they have lost four leagues of land. These sands are its harbingers; and we see great dunes of moving sand that march half a league ahead of it and keep conquering land).[49]

In an essay that in part addresses the encroachment of European physical and epistemological settlement on the indigenous inhabitants of the Americas, these comments about the loss of land to the sea are a metaphor of such settlement and the destruction it wreaks. The sands that the ocean vomits or spews, as a living creature does, are 'fourreurs': in sixteenth-century usage, literally *avant-coureurs*, the first soldiers who set up camp for an army on new territory.[50] These sands function like the soldiers who come out of the sea onto the Americas, burying the land whose conquest they begin. Montaigne's brother's land is buried in this sand—*ensevelie*, a word that principally refers to the burying of bodies. This metaphor recalls Montaigne's moral condemnation of the conquest of the Americas in the companion essay to 'Des cannibales', 'Des coches' ('Of Coaches'), Chapter 6 of Book 3, in which he laments '[t]ant de villes rasées, tant de nations exterminées, tant de millions de peuples passez fil de l'espée' ('[s]o many

48 Montaigne, *Essais* (1999), 1.31.204; Montaigne, *The Complete Essays*, 1.31.151.

49 Montaigne, *Essais* (1999), 1.31.204; Montaigne, *The Complete Essays*, 1.31.151.

50 Villey, in Montaigne, *Essais* (1999), p. 204 n. 3.

cities razed, so many nations exterminated, so many millions of people put to the sword').[51]

As I have written elsewhere, this metaphor may be taken as supplementing the discussion of New World colonization, illustrating its devastating effects.[52] But in order for it to provide this illustration, it must also suggest the converse: that the sea's erosion is akin to colonial settlement, that Montaigne attributes to the sea and sand the capacity for wilful action, extending his recognition in the 'Apologie' of nonhuman agency. In speaking of the disappearance of usable land under this settlement by 'fourriers', he proposes no remedy, although such coastal defenses as dikes, for example, had been in use in the Netherlands for several hundred years. Montaigne's omission of any response on the part of human beings or the affected land seems to suggest a passive acceptance of the water's inevitable Heraclitean change, and perhaps also a Stoic resignation to devastation. But as is evident in his moral judgment on the colonization of the Americas, he holds human beings responsible for what they do to other human beings, even if there is no turning back the clock on imperial expansion. He may be emphasizing the intractability of the sea's erosive action and its basic indifference to the destruction to human creations that it causes: just as there is little to no communication between animals and human beings, there may be little to no communication between these natural phenomena and human beings. Again, as with the 'Apologie', his point is mainly to emphasize the limitations of human knowledge and physical capacity and, in light of these limitations, the possibility that much that falls outside our awareness is itself teeming with something entirely comparable to sentient life, if it is not sentient life itself. And that suggestion is the beginning of an awareness of these life forces or lifelike forces, a point of entry into contact with them.

Conclusion

Montaigne not only recognizes the limits of human cognition and agency, thereby pointing to ways of apprehending nonhuman cognition and agency in supposedly lifeless matter, but he does so through gentle but effective challenges to habits of thinking ingrained in the Western philosophical tradition. The theoretical approach I have outlined illuminates Montaigne's strategies at the same time as it borrows from them. As I have argued, it is

51 Montaigne, *Essais* (1999), 3.6.910; Montaigne, *The Complete Essays*, 3.6.695.
52 Melehy, *Poetics*, pp. 188–191.

inevitable to impose at least some contemporary concepts on early modern texts: in addition to allowing scholars to see these texts as part of the history that precedes and informs the present day, to do so shows that theory is a practice that continually renews itself through engagements with the past, that it stems from the past as much as from the present. The application model of theory that I have criticized assumes the past as ready-to-hand for the tools of the present; that is, it regards the past as a kind of environment to be worked on and exploited. The model I develop, which has always been available in theoretical approaches to early modern literature, invites a communication with the past that challenges any notion that the present dominates it. Allowing Montaigne's writings to communicate with the present involves a humbling, a disposition that sets aside triumphalist attitudes toward the past, attitudes equally (if not more) at work in historicist approaches. The very confrontation with a past challenge to persistent habits of thinking may be part of learning the humility necessary for respecting the many lives of matter.

Works Cited

Aristotle, *The Complete Works of Aristotle: The Revised Oxford Translation*, ed. by Jonathan Barnes, 2 vols. (Princeton: Princeton University Press, 1984).

———, *Economics*, trans. by E.S. Forster, in *Complete Works*, vol. 2, pp. 2130–2151.

———, *Generation of Animals*, trans. by E. Platt, in *Complete Works*, vol. 1, pp. 1111–1218.

———, *Nichomachean Ethics*, trans. by W.D. Ross, in *Complete Works*, vol. 2, pp. 1729–1867.

———, *On the Soul*, trans. by J.A. Smith, in *Complete Works*, vol. 1, pp. 641–92.

Bennett, Jane, *Vibrant Matter: A Political Ecology of Things* (Durham, NC: Duke University Press, 2010).

Bowen, Barbara, 'Rabelais's Unreadable Books', *Renaissance Quarterly* 40.4 (1995), 742–758.

Cave, Terence, *The Cornucopian Text: Problems of Writing in the French Renaissance* (Oxford: Oxford University Press, 1979).

Conley, Tom, 'Cataparalysis', *diacritics* 8.3 (Autumn 1978), 41–59.

Cusset, François, '*Becoming Deleuzian*: Deleuze aux États-Unis, l'inconnu et la boîte à outils', in *Deleuze et les écrivains: Littérature et philosophie*, ed. by Bruno Gelas and Hervé Micolet (Nantes: Defaut, 2007), pp. 347–361.

———, *French Theory: Foucault, Derrida, Deleuze & Cie et les mutations de la vie intellectuelle aux États-Unis* (Paris: Découverte, 2005).

Dancy, R.M., *Plato's Introduction of Forms* (Cambridge: Cambridge University Press, 2004).

de Man, Paul, 'Conclusions: Walter Benjamin's "The Task of the Translator"', in *Resistance*, pp. 73–105.

———, 'The Resistance to Theory', in *Resistance*, pp. 3–20.

———, *The Resistance to Theory* (Minneapolis: University of Minnesota Press, 1986).

Defaux, Gérard, *Marot, Rabelais, Montaigne: L'écriture comme présence* (Paris and Geneva: Champion-Slatkine, 1987).

———, Review of Floyd Gray, *Rabelais et l'écriture*, *The French Review* 48.6 (May 1975), 1049–1050.

Deleuze, Gilles *Différence et répétition* (Paris: Presses Universitaires de France, 1968).

———, *The Logic of Sense*, trans. by Mark Lester with Charles Stivale, ed. by Constantin Boundas (New York: Columbia University Press, 1990).

———, *Logique du sens* (Paris: Minuit, 1969).

———, 'Lucrèce et le simulacre', in *Logique du sens*, pp. 307–324.

———, 'Lucretius and the Simulacrum', in *The Logic of Sense*, pp. 266–279

———, 'Plato and the Simulacrum', in *The Logic of Sense*, pp. 253–266

———, 'Platon et le simulacre', in *Logique du sens*, pp. 292–307.

Derrida, Jacques, 'L'animal que donc je suis (à suivre)', in *L'Animal autobiographique: Autour de Jacques Derrida*, ed. by Marie-Louise Mallet (Paris: Galilée, 1999).

———, 'The Animal That Therefore I Am (More to Follow)', in *The Animal That Therefore I Am*, ed. by Marie-Louise Mallet, trans. by David Wills (New York: Fordham University Press, 2008), pp. 1–51.

Descartes, René, letter to the Marquess of Newcastle, 23 November 1646, in *The Philosophical Writings of Descartes*, trans. by John Cottingham, Robert Stoothoff, Dugald Murdoch, and Anthony Kenny, 3 vols. (Cambridge: Cambridge University Press, 1984–1991), vol. 3, pp. 302–304.

Harari, Josué, 'Critical Factions / Critical Fictions', in *Textual Strategies: Perspectives in Post-Structuralist Criticism*, ed. by Harari, Josué (Ithaca: Cornell University Press, 1979), pp. 17–72.

Hardt, Michael, *Gilles Deleuze: An Apprenticeship in Philosophy* (Minneapolis: University of Minnesota Press, 1993).

Hayot, Eric, 'Against Historicist Fundamentalism', *PMLA* 131.5 (2016), 1414–1422.

Irwin, T.H., 'The Theory of Forms', in *Plato 1: Metaphysics and Epistemology*, ed. by Gail Fine (Oxford: Oxford University Press, 1999).

Krause, Virginia, *Idle Pursuits: Literature and* Oisiveté *in the French Renaissance* (Newark, DE: University of Delaware Press, 2003).

Lucretius, *On the Nature of the Universe*, trans. by R.E. Latham (London: Penguin, 1951).

Mathieu-Castellani, Gisèle, *Montaigne, l'écriture de l'essai* (Paris: Presses Universitaires de France, 1988).
Melehy, Hassan, 'Images Without: Deleuzian Becoming, Science Fiction Cinema in the Eighties', *Postmodern Culture* 5.2 (January 1995): n.p.
———, *The Poetics of Literary Transfer* (Burlington, VT and Farnham, UK: Ashgate, 2010).
———, 'Silencing the Animals: Montaigne, Descartes, and the Hyperbole of Reason', *symplokē* 13.1–2 (2005), 263–282.
Montaigne, Michel de, *The Complete Essays of Montaigne*, trans. by Donald M. Frame (Stanford, CA: Stanford University Press, 1958).
———, *Essais*, ed. by Pierre Villey (Paris: Presses Universitaires de France, 1999), 3 vols.
———, *Les Essais* (Paris: Abel L'Angelier, 1595), 1.8.16–17, Gallica digitized edition.
———, *Œuvres complètes*, ed. by Albert Thibaudet and Maurice Rat (Paris: Gallimard, 1963).
Nealon, Jeffrey, 'Exteriority and Appropriation: Foucault, Derrida, and the Discipline of Literary Criticism', *Cultural Critique* 21 (Spring 1992), 97–119.
O'Brien, John, 'Introduction: The Time of Theory', in *Distant Voices Still Heard*, ed. by O'Brien and Malcolm Quainton (Liverpool: Liverpool University Press, 2000), pp. 1–52.
Plato, *The Collected Dialogues, Including the Letters*, ed. by Edith Hamilton and Huntington Cairns (New York: Pantheon, 1961).
———, *Laws*, trans. by A.E. Taylor, in *Collected Dialogues*, pp. 1225–1513.
———, *Phaedo*, trans. by Hugh Tredennick, in *Collected Dialogues*, pp. 40–98.
———, *Phaedrus*, trans. by R. Hackforth, in *Collected Dialogues*, pp. 475–525.
Plutarch, *La Mesnagerie de Xenophon, Les Regles de mariage de Plutarque*, trans. by Estienne De la Boétie (Paris: Frédéric Morel, 1571), Gallica digitized edition.
———, *Les Œuvres morales de Plutarque*, trans. by Jacques Amyot (Paris: Michel de Vascosan, 1574).
Regosin, Richard, *Montaigne's Unruly Brood: Textual Engendering and the Challenge to Paternal Authority* (Berkeley: University of California Press, 1996).
Ricœur, Paul, *Being, Essence and Substance in Plato and Aristotle: Course taught at the University of Strasbourg in 1953–1954*, trans. by David Pellauer and John Starkey (Cambridge: Polity Press, 2013).
Ross, David, *Aristotle* (London: Methuen, 1923).
Sallis, John, *Being and Logos: The Way of Platonic Dialogue* (Atlantic Highlands, NJ: Humanities Press International, 1986).
Screech, M.A., *Montaigne's Annotated Copy of Lucretius: A Transcription and Study of the Manuscript, Notes, and Pen-Marks* (Geneva: Droz, 1998).

Worth-Stylianou, Valerie, 'Exchanges between Medical and Literary Discourses in the French Renaissance: Comparing the Fascination with "des amas et pieces de chair informes" in Montaigne's *Essais* with Joubert's *Erreurs populaires*', *Australian Journal of French Studies* 52.3 (2015), 290–305.

About the Author

Hassan Melehy (Professor of French, University of North Carolina) currently researches the relationship between political writing and literature in sixteenth- and seventeenth-century France and England, developing the themes of his 2010 book, *The Poetics of Literary Transfer in Early Modern France and England* (Ashgate), and the collection he recently co-edited with Catherine Gimelli Martin, *French Connections in the English Renaissance* (Ashgate). He has also written numerous articles on early modern literature and philosophy, recent and contemporary critical theory, and film studies.

Part 1

Dark(Ish) Ecologies

2. Du Bartas Responding to Morton's Milton: A Bodily Route to the Ecological Thought

Stephanie Shiflett

Abstract
In *The Ecological Thought* (2012), Timothy Morton calls us to recognize the interconnectedness of all things by rethinking the relationship between cosmic and local. He points to Raphael's speech to Adam in John Milton's *Paradise Lost*, which compares Earth to the infinite cosmos, as an example of this ecological thought. An analogous cosmic viewpoint occurs in Guillaume du Bartas's *La Sepmaine* (1578). This hexameron both highlights and complicates ecocriticism's applicability to early modern texts. Whereas Milton's text responds to Morton's call by scaling Earth in relation to the macrocosmic, Du Bartas's does the opposite: it scales the cosmic to the hyperlocal—the observer's body. This earlier work thus offers a converse avenue by which to arrive at the ecological thought.

Keywords: Lucretius, dark ecology, macrocosm, senses, hexameron

If the purpose of this volume is to tease out the forms of ecology present in early modern France, then it seems we have our work cut out for us. First, what do we mean by 'ecology'? Second, how does one even begin to look for this ecology in an era far removed from our own, where the concept risks anachronism? Timothy Morton modelled a possible means of addressing these challenges in his 2010 book, *The Ecological Thought*. Morton's title refers to a perspective on the natural world that he advocates in the book, and which I will adopt for the purposes of this essay. It is not unusual for Morton to use early modern texts to explain his theories. For example, he uses John Milton's *Paradise Lost* (1667), a work that draws deeply on

Goul, P. and P.J. Usher (eds.), *Early Modern Écologies. Beyond English Ecocriticism.*
Taylor & Francis Group, 2020
DOI 10.5117/9789462985971_CH02

Renaissance sources, to illustrate the ecological thought. I wish to extend Morton's technique to one of Milton's primary inspirations and one of the most popular authors of the late sixteenth century, the French Protestant Guillaume du Bartas (1544–1590). In the following chapter, I will show how Du Bartas arrives at this ecological point of view in a different way from Milton. Whereas Milton employs a perspective that zooms outward from Earth, Du Bartas zooms inward, into the human body, in order to offer up a differently inflected view of the ecological thought.

Morton's Ecological Thought

Timothy Morton's recent works, including *The Ecological Thought* and its predecessor, *Ecology without Nature,* seek to inject theory into ecological criticism. Morton claims that theory—specifically, deconstruction—is necessary to truly shift society's thinking about the environment in a way that will compel us to treat it better. *Ecology without Nature* (2007) deconstructs the idea of 'nature' created in the Romantic age, which still persists today, and which Morton claims keeps the physical world *over there*. In *The Ecological Thought,* meanwhile, Morton calls upon us to think big—cosmically big. He holds up Tibetan Buddhists as a model, for whom the universe, and billions of other universes, are a speck inside of a lotus flower held by an immense Buddha.[1] This image is integral to the necessary shift in perspective that must occur in order to respond to the global climate crisis. Morton's 'ecological thought' additionally breaks down the barrier between human and nonhuman animal. It is not enough to think that humans are animals or that animals are human-like; one must accept that the categories 'human' and 'animal' are fallacious. As he undoes the category of 'animal', he replaces it with what he calls the 'strange stranger': a being related to us, uncannily like us, if not us (a nod to Derrida's *arrivant*).

The interconnectedness of all things, of all strange strangers, Morton calls 'the mesh'. By his account, the perspective of the ecological thought allows us to fully realize how interconnected we are with the natural world, and thus realize that harming the environment means harming ourselves. Rather than a trite platitude, Morton's injunction is critically—radically—pessimistic. The environment has already deteriorated, and has taken us down with it. By espousing this pessimistic view of global warming, Milton offers a response

1 Morton, *The Ecological Thought,* p. 26.

to a problem frequently encountered and debated within the community of climate scientists and commentators: How can one talk about the climate crisis to the general population in a way that will motivate them to act, or as Morton might suggest, cope?[2]

Morton's rather blunt approach of instilling fear of global warming in his readers is controversial.[3] Elizabeth Boulton, for example, argues that this approach may leave humans feeling so powerless that they refuse to respond to the climate crisis at all.[4] However, in my view, it is precisely this apocalyptic pessimism that makes the ecological thought so appealing, and so relevant to the quest of recognizing a sixteenth-century ecology. Frenchmen of the sixteenth century were not trying to stop the end of the known world, as tends to be the approach of ecological activists today; rather, the end of the known world had already happened, for all intents and purposes. The sixteenth century was a time of rapid epistemological change: geographical discoveries were challenging cosmological and anthropological beliefs, and Europe saw an explosion of mapmaking as intellectuals struggled to reformulate their view of the cosmos.[5] Whereas medieval man had had no reason to doubt an objective reality described by the likes of Aristotle and Ptolemy, and of course, the Bible, intellectuals of the Renaissance had to contend with tensions between received wisdom and new discoveries. What to make of the tripartite maps that divide global space into the three continents of Europe, Asia, and Africa after the European discovery of the New World?[6] What to make of Galen after Vesalius disproved the former's statement that blood passed through the septum of the heart? What to make of taking the Eucharist when theologians could not agree on what it actually meant? These doubts did not immediately undermine the authority of received wisdom, but chipped away at its foundation until it finally crumbled to pave the way for the scientific revolution. A late sixteenth-century French ecology would have to negotiate an ontological apocalypse that had already happened, which makes Morton so applicable here.

2 See Boulton, 'Climate Change as a "Hyperobject"'.

3 This is by no means the only controversial aspect of Morton's work. His rapid, catchy writing style can be enormously entertaining, but irritates many who claim that it causes him to pass too quickly over nuances and definitions. A number of reviewers have succinctly captured the controversies and debates surrounding Morton's work, notably Mick Smith, 'Dark Ecology'.

4 See Boulton, 'Climate Change at a "Hyperobject"'.

5 See Lestringant, *L'Atelier du cosmographe*.

6 See Usher, 'The Holy Lands in Early Modern Literature' and Masse, 'Newness and Discovery in Early Modern France'.

In *The Ecological Thought,* Morton draws on Milton to illustrate the sort of uber-macrocosmic thinking that characterizes the ecological thought. He cites Raphael's speech to Adam in Book Seven of *Paradise Lost*:

> What if that light
> Sent from her through the wide transpicuous air,
> To the terrestrial moon be as a star
> Enlight'ning her by day, as she by night
> This earth? reciprocal, if land be there,
> Fields and inhabitants: her spots thou seest
> As clouds, and clouds may rain, and rain produce
> Fruits in her softened soil, for some to eat
> Allotted there; and other suns perhaps
> With their attendant moons, thou wilt descry
> Communicating male and female light,
> Which two great sexes animate the world,
> Stored in each orb perhaps with some that live.
> For such vast room in Nature unpossessed
> By living soil, desért and desolate,
> Only to shine, yet scarce to contribute
> Each orb a glimpse of light, conveyed so far
> Down to this habitable, which returns
> Light back to them, is oblivious to dispute. (VIII. 140–158)

Morton hints that there is something subversive about Raphael's speech: at the same time as he warns Adam against flights of fancy, he takes him on a fantastical thought-journey through outer space. The point, ostensibly, is not to waste time with idle thoughts, but to take into account our own smallness in relation to the universe, to not let our egos get puffed up thinking that we are the only inhabitants here. As Morton points out, Milton uses this moralizing tactic in Book III of *Paradise Lost* to make Satan, the biblical embodiment of hubris, look small.[7] Morton himself is playing the role of Milton with his work. By 'humbling' the human, Morton addresses the issue of hubris, identified by climate policy analysts as inhibiting adequate responses to global warming.[8] Morton compares this to the proverbial injunction not to think of a pink elephant: once stated, the commanded

7 Morton, *The Ecological Thought,* p. 23.
8 See Boulton, 'Climate Change as a "Hyperobject"', p. 777.

has already thought of a pink elephant, and the injunction is moot.[9] Morton highlights Raphael's thought experiment as an example of the ecological thought. It incites one to think big, on a Buddhist scale, and acknowledges the 'mesh' created by a connection to other Adams and Raphaels on other planets. Morton writes:

> Raphael doesn't claim that extraterrestrials exist: that's the whole point. The mere possibility of extra-terrestrial environments and sentient beings—their possibility (hypothetical but imperceptible) is their essence—provides the fantasy point from which the reader herself, like Adam or Eve, can achieve the 'impossible' viewpoint of space. To reach this standpoint involves an act of rational self-reflection independent of graven images. This 'impossible' viewpoint is a cornerstone of the ecological thought.[10]

This kind of thought journey entails intellectual travel to a viewing point that is both cosmic and nowhere, through mental zoom lenses that, like 'the opening sequence of the film *Contact*, based on Carl Sagan's novel, travel out, and out, and out, from Earth into the Universe'.[11] Even if the passage from Milton is not exactly parallel with the opening sequence of *Contact*—Raphael's mental perspective can be localized *on* the Earth and *on* the moon, which is not entirely nowhere—we can still imagine Raphael and Adam's thoughts zooming out from their place on Earth, to the moon and beyond.

The Ecological Thought in *La Sepmaine*

A cosmic viewpoint comparable to the one that Raphael shows Adam in *Paradise Lost* occurs at the beginning of the seventh book of Du Bartas's *La Sepmaine*. The title, *La Sepmaine,* refers to the biblical week in which God created the world, according to the Book of Genesis. The work was immediately popular. *La Sepmaine* is divided into seven parts, each of which corresponds to a day of creation. The book is more than just a

9 Morton, *Ecological Thought,* p. 21.

10 Ibid., pp. 22–23.

11 Morton, *The Ecological Thought,* p. 24. On this impossible 'view-from-nowhere' point of view, and Bruno Latour's rejection thereof, see Usher, 'The Revenge of the Mines: Earth-from-Nowhere versus Surfaces-with-Depths'.

commentary on Genesis, however.[12] Du Bartas goes into encyclopaedic detail about the natural wonders of the world, from stars to bodily humours to cuttlefish.

The seventh book corresponds to the seventh day in the Book of Genesis, on which God rested and admired his work (Gen. 2.2). To recall, in the passage of interest in *Paradise Lost*, Raphael stands on Earth with Adam and invites him to think of possible inhabitants on the moon and beyond. In Du Bartas's account, God looks at his handiwork, pleased with the result, and the rest of the passage lists the different details of creation that he sees there. We, the readers, look through God's eyes toward Earth. In the first verses of the seventh book, Du Bartas compares God to a painter who steps back to admire his own masterpiece:

Le Peintre qui, tirant un divers paysage,
A mis en œuvre l'art, la nature, et l'usage,
Et qui d'un las pinceau sur si docte pourtrait
A, pour s'eternizer, donné le dernier traict:
Oublie ses travaux, rit d'aise en son courage,
Et tient tousjours ses yeux collez sur son ouvrage. (VII. 1–6)[13]

(The cunning *Painter*, that with curious care,
Limning a Land-scape, various, rich, and rare,
Hath set a-work, in all and every part,
Invention, judgment, Nature, Use, and Art;
And hath at length (t'immortalize his name)
With weary Pencill perfected the same;
Forgets his pains; and, inly fill'd with glee,
Still on his *Picture* gazeth greedily). (VII. 11–18)[14]

12 For information on the revival of commentaries of Genesis in the sixteenth century, see Banderier, 'Un "Heureux Phénix"? Renaissance et mort de l'hexaméron (1578–1615)' and Williams, 'Commentaries on Genesis as a Basis for Hexaemeral Material in the Literature of the Late Renaissance.'

13 The comparison between God and a painter is philosophically significant. The world as painting suggests that the world is illusory, a hallucinatory by-product of the real cosmic forces at work. Du Bartas reiterates this typically early modern belief by referring to the world as 'un théâtre' and 'un grand livre'. We will return later in this essay to the importance of sensory organs in Du Bartas's depiction of the creation of the world.

14 I am using Josuah Sylvester's 1621 translation of the *Divines Semaines*, which remains the most widely-used translation into English of Du Bartas's work. However, Sylvester takes significant liberties with his translation, and I have eliminated some of his insertions for the sake of clarity.

In Milton's work, Raphael and Adam stand on a fixed point, in the Garden of Eden, and look outward. In this passage from Du Bartas, however, God looks to the earth from ... where? We cannot know where Du Bartas's God is standing when he is looking at the world. Logic would seem to dictate that God looks down from a point in the heavens, but in fact, this is not stated in the text. God's actual location in this passage is not of concern to Du Bartas. According to the text, God looks down at the world from the nowhere point of view described by Morton, what the theorist describes as the ambiguous position of the photographer of 'Earthrise'.[15]

To reach this nowhere point of view, looking down at Earth from outside of it, the mind must travel. Morton does not necessarily emphasize the importance of travel to the ecological thought, but it is clearly implied by his explanation of the nowhere point of view. While elaborating on the implications of this concept, Morton rhetorically asks: 'Do we have to go into outer space to care for Earth? Do we need high technology?' and then answers simply, 'No'.[16] His point is that the nowhere point of view must be achieved through a shift in mental perspective. In other words, we have to be able to travel intellectually to this zoomed-out point of view in order to engage in the ecological thought. Just as Raphael entices Adam's thoughts further and further away from Earth, Du Bartas takes his reader on an intellectual voyage through the cosmos in *La Sepmaine*.

Du Bartas describes the intellectual journey on which he leads the reader in terms of sailing and navigation. This is a reminder that Du Bartas's text (and by consequence, the form of ecological thought present in that text) is a product of the so-called Age of Exploration. Du Bartas seems to be aware that his text has conceptual links to European voyages. He repeatedly alludes to the concrete realities of naval exploration, and even to specific explorers. For example, Du Bartas argues that the world is round, taking as evidenced the discoveries of Vespucci, Columbus, and Marco Polo.[17]

> Voire quand un Vespuce, un Colomb, un Marc Pole,
> Et cent autres Typhis n'auroyent sous autre pole
> Conduit le pole arctique, et vivans sur les eaux

15 For an analysis of related questions, see Usher, 'The Revenge of the Mines: Earth-from-Nowhere versus Surfaces-with-Depths'.

16 Morton, *The Ecological Thought*, p. 25.

17 On Du Bartas's mapping of the New World, see *inter alia*, Usher, *L'Aède et le géographe*, pp. 306–316.

Trouvé dessous nos pieds tant de mondes nouveaux. (III. 365–372)
Yea, though *Vespasio* (famous *Florentine*)
Marke Pole, and *Columb*, brave *Italian* Trine,
. . . .
And thousand gallant modern *Typheis* else,
Had never brought the *North-Poles* Parallels
Under the *South*; and, sayling still about,
So many *New-worlds* under us found out. (III. 421–432)

We can, perhaps, see a link between Du Bartas's evocation of the New World and Raphael's hypothesis of inhabited worlds beyond Earth. The possibility of worlds beyond our own—which Morton asserts is central to Raphael's demonstration of the ecological thought—became a very real possibility after the voyages of the explorers that Du Bartas refers to in this passage.

Du Bartas even presents his text as travel literature. With the opening lines of the sixth book (which corresponds to Day Six, or the creation of man), Du Bartas invites his reader on a literary pilgrimage:

Pelerins, qui passez par la cité du monde,
Pour gagner la cité, qui, bien-heureuse, abonde
En plaisirs eternels, et pour ancrer au port,
D'où n'aprochent jamais les horreurs de la mort:
Si vous desirez voir les beaux amphitheatres,
Les arsenals, les arcs, les temples, les theatres,
Les colosses, les ports, les cirques, les rempars,
Qu'on void superbement dans nostre ville espars,
Venez avecque moy. (VI. 1–9)

(You Pilgrims, which (through this worlds Citie) wned
Toward th'happy Citie, where withouten end
True joyes abound; to anchor in the Port
Where Deaths pale horrors never do resort?
If you will see the fair Ampitheaters,
Th'Arks, Arcenals, Towrs, Temples, and Theaters,
Colosses, Cirques, Pyles, Ports, and Palaces
Proudly dispersed in your Passages;
Com, com with me: for, there's not any part
In this great Frame where shineth any Art,
But I will show't you). (VI. 11–21)

Du Bartas invites the reader on a journey toward a marvelous city, which clearly resembles ancient Rome. The destination is 'D'où n'aprochent jamais les horreurs de la mort': the written word that paves the way of this voyage ensures immortality, both of the artist through his works and of the ideas materially encoded in the text.

Yvonne Bellenger notes the irony of a firm Calvinist writing about the very Catholic habit of pilgrimage. She comments on these first lines of Book Six: 'On aurait tort de s'étonner ou de sourire devant cette allusion aux pèlerinages sous la plume d'un calviniste, alors que chacun sait combien les réformés se montraient hostiles à ces pratiques de dévotion superstitieuse. Car l'interpellation est ici purement oratoire, de même qu'est purement symbolique le pèlerinage, et aussi le "port"'[18] ('It would be wrong to be surprised or to smile before this allusion to pilgrimages from a Calvinist's pen, since how hostile reformists showed themselves toward this superstitious practice of devotion is well-known. Because the interpellation here is purely rhetorical, and also the pilgrimage is purely symbolic, and also the "port"'). But is dismissing the reference to pilgrimage as simply a rhetorical device, totally separate from the author's worldview, too facile? Elsewhere in *La Sepmaine,* Du Bartas describes intellectual contemplation as a sort of extra-corporeal journey of the spirit, which is also a form of pilgrimage:

Car quittant quelquefois les terres trop connues,
D'une alegre secousse il saute sur les nues,
Il noue par les airs, où, subtil, il aprend
Dequoy se fait la neige, et la gresle, et le vent:
. . . .
Il conte leurs brandons, il mesure leurs pas,
Il aulne leur distance: et comme si le monde
N'enfermoit dans le clos de sa figure ronde
Des subjets assez beaux, il s'eslance dehors
Les murs de l'Univers: et loin, loin de tous corps
Il void Dieux face à face, il void les chastes gestes
Et le zele fervent des courtisans celestes. (VI. 789–812)

(For, sometimes, leaving these base slimy heaps,
With cheerfull spring above the Clouds she leaps,
Glides through the Aire, and there she learns to knowe

18 Bellenger, *Du Bartas et ses Divines Semaines,* p. 73.

Th'Orginals of Winde, and Hail, and Snowe,
. . . .
She counts their Stars, she metes their distances
And differing pases; and, as if she found
No Subject fair enough in all this Round,
She mounts above the Worlds extreamest Wall,
Far, far beyond all things corporeall;
Where she beholds her Maker, face to face). (VI. 847–863)

Du Bartas describes the spirit leaving the body, flying up to the skies, learning about the natural world in the process, then up to Heaven, where it comes face-to-face with God. He may not simply be using travel as a metaphor here. According to Dudley Wilson, Du Bartas would not have been alone in conceiving of intellectual pursuit as a form of spiritual travel in a quite literal sense. Wilson mentions 'the seven ways of detaching the soul from the body in its pursuit of mystic contemplation' as enumerated by Lefèvre de la Boderie, another French scientific poet[19] and contemporary of Du Bartas.[20] Among Du Bartas's literary contemporaries, specifically the scientific poets, spiritual travel as a means of mystical union with God was an idea with currency. Du Bartas's text, then, has this in common with the passage of Milton's cited by Morton: travel, even imaginary travel, is necessary to imagine other bodies, other worlds, which is integral to the ecological thought.

Journey to the Ecological Thought

In Book Seven, at the end of this intellectual voyage, the reader stands with the character of God from the nowhere point of view discussed above. Du Bartas's inclusion of the seventh day emphasizes the importance of the nowhere perspective to his text. This day is unique to Du Bartas: previous hexamera, including the prime examples of Basil and Ambrose, had only treated the first six days (hence the name of the genre, *hex*ameron). This seventh day is important, then, in that it is a key element of what is unique in Du Bartas's work, and not just an imitation of early hexamera and commentaries on Genesis.[21]

19 Due to Du Bartas's particularly thorough admiration of nature, scholars have long referred to his work as 'scientific poetry'. Albert-Marie Schmidt made this categorization standard with his 1938 study, *La poésie scientifique en France au seizième siècle*.

20 Wilson, *French Renaissance Scientific Poetry*, p. 4.

21 For more on Du Bartas and the hexameral tradition, see Reichenberger, 'Das Schöpfungsepos'.

In short, Du Bartas seems to include the seventh book of *La Sepmaine* to showcase a God-like visual perspective on the world. The beginning of Book Seven recounts what God sees when he looks down from nowhere:

> Il regarde tantost par un pré sauteler
> Un agneau, qui tousjours, muet, semble besler.
> Il contemple tantost les arbres d'un bocage,
> Ore le ventre creux d'une grotte sauvage,
> . . . .
> Ici fume un chasteau, là fume une cité :
> Et là flote une nef sur Neptune irrité. (VII. 7–40)

> (First, in a Mead he marks a frisking Lamb,
> Which seems (though dumb) to bleat unto the Dam:
> Then he observes a Wood, seeming to wave:
> Then th'hollow bosom of som hideous Cave:
> . . . .
> Heer smokes a Castle, there a Citie fumes,
> And heer a Ship upon th'Ocean looms). (VII. 19–54)

God looks down at a pastoral hunting scene, like a European king admiring a tapestry in his hunting lodge. Du Bartas uses this point of view to list different features of the natural world in a copious litany of awe. In this passage, as throughout *La Sepmaine,* different features of the world temporarily monopolize God's vision. Each is allowed a brief moment as protagonist. A city earns the same amount of attention as a single felled oak tree. Humans, animals, and plants are placed on equal levels—they each occupy God's vision, are the protagonists of the Earth story, for a brief moment at a time. This approach levels the playing field of different beings, defamiliarizing our sense of scale.

This defamiliarization of scale is precisely what Morton calls for in *The Ecological Thought.* He would have us, like Adam, realize how small we are in the greater scheme of things; at the same time, we must magnify the features of the natural world that humans have long suppressed as less important than ourselves, realizing that we are all on equal footing in this great interconnected mesh. This defamiliarization of scale harkens to Morton's background of object-oriented ontology that, as Boulton summarizes simply, 'asserts not only that humans and nonhuman objects have different experiences of "Being," but that humans have equal existential status to other "objects," such as frogs, chairs, or planets'.[22] Du Bartas achieves the defamiliarization of scale that forms a cornerstone of the ecological thought throughout *La Sepmaine.* He spends a seemingly inordinate numbers of verses on nonhuman creatures, with a result that the cuttlefish, with twenty-five lines, becomes a significant figure of Book Five.

While Du Bartas gives nonhuman creatures their own ontological status, he does not suggest that they are man's equal. Humans retain their biblical place as lord of other species. Building on Schmidt's observation to this effect, Yvonne Bellenger sees in *La Sepmaine* 'le désir plus ou moins conscient de répéter l'acte d'Adam, le premier ancêtre que Dieu avant la chute avait invité à nommer le monde' ('the more or less conscious desire to repeat the act of Adam, the first ancestor whom God invited to name the world before the fall').[23] It is true that Du Bartas is far from modern object-oriented ontology. He is constrained by Aristotelian philosophy and its 'Great Chain of Being', as well as the letter of Genesis.[24] However, the attention that he gives to nonhuman beings, and the agency he gives them by making them the protagonists of their own vignettes, anticipates Morton's goals: nonhumans become 'strange strangers' to humankind, uncannily similar but other, and a conventional sense of scale between large creatures and small loses its footing.

Sense as Medium

We move now from Du Bartas's treatment of the objects of his attention to the relationship between object and observing subject. In *The Ecological Thought,* Morton writes:

22 Boulton, 'Climate Change as a "Hyperobject"', p. 776.

23 Bellenger, 'L'Intelligence des animaux', p. 534. Hélène Naïs also addressed this question in *Les Animaux dans la poésie française de la Renaissance,* concluding: 'C'est dire que, si l'homme partage avec les animaux la destinée de créature, il reste toujours une créature supérieure, même après la chute. En cela, on peut dire que Du Bartas fournit l'expression la plus complète de l'opinion chrétienne sur le problème' (That is to say, if man shares with animals the fate of a creature, he remains a superior creature, even after the fall. In this we can say that Du Bartas provides the most complete expression of the Christian opinion on the matter' [p. 562]).

24 Primarily, Gen. 1. 26: 'And God said, Let us make man in our Image, after our likenesse: and let them have dominion over the fish of the sea, and over the foule of the air, and over the cattell, and over all the earth, and over every creeping thing that creepeth upon the earth.' (King James Version).

> The ecological thought reserves a special place for the 'subject'—the mind, the person, even the soul. Posthumanism seems suspiciously keen to delete the paradigm of humanness like a bad draft [...]. What if being human is the encounter with the strange stranger—in other words, at a certain limit, an encounter with the inhuman? Isn't this the very 'posthumanism' for which some are yearning?[25]

Morton does not want to do away with the human subject, but rather define it by its interactions with other beings. Du Bartas obliges with this concept of the human subject. For Du Bartas, the observing subject—whether human or divine—is present mainly in the form of an anthropomorphic body. That body is defined by its physical interaction with other organisms, be they animal or mineral, via the sensory organs. Recall the first verses of Book Seven, in which God, like a painter admiring his masterpiece, 'tient tousjours ses yeux collez sur son ouvrage' (VII.6) ('Still on his *Picture* gazeth greedily'): this passage emphasizes God's sense of sight, and specifically, his eyes. This emphasis on sensory organs, both humanity's and God's, pervades *La Sepmaine*. Later on in Book Seven, after listing what God sees on earth while he rests after creating it, Du Bartas writes:

> Et bref, l'oreille, l'œil, le nez du Tout–puissant,
> En son œuvre n'oit rien, rien ne void, rien ne sent,
> Qui ne presche son los, où ne luise sa face,
> Qui n'espande par tout les odeurs de sa grace.
> Mais plus que tous encor les humaines beautez
> Tienent du Tout–puissant tous les sens arrestez. (VII. 91–96)
>
> (In briefe, th'Almightie's *ey,* and *nose,* and *ear,*
> In all his works, doth nought *see, sent,* or *hear*
> But *showes* his greatness, *savours* of his grace
> And *sounds* his glory over every place.
> But above all, Mans many beautious features
> Detaine the Lord more than all other Creatures). (VII. 105–110)

God's anthropomorphic ears, eyes, and nose are enthralled by the sensory information emanating from his creation. Like humans, God interacts with the world through the medium of senses. This concept

25 Morton, *The Ecological Thought,* p. 113.

of sensory interaction between the subject and its environment comes from Lucretius, a poet whom Du Bartas explicitly rejects, but to which he is implicitly indebted.[26] As scholars such as Stéphane Lamacz have pointed out, Du Bartas takes up Lucretius's position that the natural world can be known through its effects on the body. In other words, any broad perspective must be constituted through the corporeal subject. Lucretius posited that objects in the world emitted 'simulacra' which are physically absorbed into the human body, in the form of atoms, by sensory organs.[27]

The poem itself, by stimulating the reader's senses through imagery and sound, acts as a physical body affecting the reader's senses.[28] Through its sonority, as well as its visual presence on the page, Du Bartas's text physically influences the reader's sensory organs. In other words, Du Bartas's text is a body. The text-body trope is centuries old at least, but finds new life in Morton's writings. In an article called 'Ecology as Text, Text as Ecology', which expands on some of the ideas presented in *The Ecological Thought,* Morton elaborates on his idea of text as a concrete entity. He writes:

> Text as ecology is a good metaphor. But thinking can go much further than this, since if the text has no thin, rigid boundary, what it includes, what it touches, must also consist of life forms, Earth itself, and so on. The difference between what counts as a mere metaphor and what counts as non-metaphorical reality collapses when thinking engages text seriously.[29]

In Morton's argument, text is a *thing,* inextricably bound with everything involved in the conditions of its being that, by a sort of butterfly effect, includes everything. For Du Bartas, as well as for other Renaissance intellectuals influenced by Lucretius (including John Milton),[30] text had the power to physically affect the body through the senses, and thus become part of the body itself. The boundary between body and environment was thin indeed. We are far, however, from rejecting the human subject, which Morton laments as a feature of contemporary posthumanist ideas. Body,

26 See also Lamacz, 'La Construction du savoir et la réécriture du De Rerum Natura dans La Sepmaine de Du Bartas' and Kany-Turpin, 'Une Réinvention de Lucrèce par Guillaume du Bartas'.

27 Ford, 'Lucretius in Early Modern France', p. 236.

28 Lamascz, 'La Construction du savoir', p. 631.

29 Morton, 'Ecology as Text', pp. 2–3.

30 For analysis of Lucretian influence on *Paradise Lost,* see, for example: Hardie, 'The Presence of Lucretius in Paradise Lost'.

environment, and text are three separate entities in sixteenth-century thought, but they interact in a concrete, non-metaphorical way, inextricably linked and mutually defining.

Given this Lucretian influence, it is not surprising that Du Bartas should turn to the human body in his quest to relate knowledge of the natural world to his readers.[31] The idea that the body can serve as a route to an awareness of the cosmos may thus be due in part to the influence of Lucretius. This examination of Du Bartas's use of the body as the primary means of knowing the natural world sets us up to explore how Du Bartas treatment of the body provides a vehicle for arriving at the ecological thought.

Voyage Through the Cosmic Body

We saw earlier in this essay how Du Bartas takes his readers on a voyage through the natural world, a mental journey that resembles Raphael and Adam's quest to imagine worlds beyond Earth. In this section, we will examine Du Bartas's intellectual voyage through, indeed *into*, the human body, which he often conflates with a voyage through the cosmos. In a notable passage from Book Six, on the creation of man, Du Bartas writes:

> Hé! quoy? n'est il pas temps, n'est il pas temps de voir
> Dans les secrets du corps le non-secret pouvoir
> D'un si parfait Ouvrier? Prendray-je la scalpelle
> Pour voir les cabinets de la double cervelle,
> . . . .
> Pourray-je desployer sur un docte fueillet
> Ce Dedale subtil, cest admirable reth
> Par les replis duquel l'esprit monte et devale,
> Rendant sa faculté de vitale, Animale:
> Tout ainsi que le sang et les esprits errans
> Par le chemin courbé des vaisseaux preparans
> D'un cours entortillé s'elabourent, se cuisent,
> Et en sperme fecond peu à peu se reduisent? (VI. 641–660)

> (But, is't not time now, in his Inner Parts,
> To see th'Almightie's admirable Arts?
> First, with my Launcet shall I make incision,

31 Du Bartas shares this influence with Milton.

To see the Cells of the twin Brains division:
. . . .
O, how shall I on learned Leaf forth-set
That curious Maze, that admirable Net,
Through whose fine folds the spirit doth rise and fall,
Making its powrs of *Vital, Animal!*
Even as the Blood, and Spirits, wandering
Through the *preparing vessels* crooked Ring,
Are in their winding course concoct and wrought,
And by degrees to fruitfull *Seed* are brought). (VI. 675–694)

Du Bartas continues down to the heart, then the lungs, then the stomach. The idea of travelling through the body is announced first by temporal progression: Du Bartas asks, 'n'est-il pas temps'—isn't it time—and then uses the future tense, 'Prendray-je la scalpelle' ('First, with my Launcet shall I make incision'), which he then, verbally, proceeds to do. The temporal progression here is not uniform in the text—at other points, time stands still and cedes to pure description, or even jumps backward—and announces the spatial movement through the body. A number of key, movement-related words signal to the reader the idea that we are travelling: 'Dedale', or labyrinth; how blood 'errans | Par le chemin courbé' ('wandering | Through the *preparing vessels* crooked Ring'), and spreads out through 'un cours entortillé' ('winding course') until it finally reduces, following Galenic logic, into sperm.

Du Bartas draws parallels between the functioning of the natural world and the functioning of the human body by way of the microcosm/macrocosm trope.[32] For example, in the first few pages of *La Sepmaine,* Du Bartas describes the creation of the world in terms of gestation:

Estoit tel que la chair, qui s'engendre, difforme,
Au ventre maternel, et par temps toutesfois,
Se change en front, en yeux, en nez, en bouche, en doigts
Prend ici forme longue, ici large, ici ronde,
Et de soy peu à peu fait naistre un petit monde. (I.263–268)

(for that huge lump was like
The shape-less burthen in the Mothers womb,

32 For more on the microcosm/macrocosm trope in the early modern world, see E.M. Tillyard, *The Elizabethan World Picture*, p. 92.

> Which yet in Time doth into fashion com:
> Eyes, eares, and nose, mouth, fingers, hands, and feet,
> And every member in proportion meet;
> Round, large, and long, there of it selfe it thrives,
> And (*Little-World*) into the World arrives). (I.298–304)

The forming universe in this analogy is like a foetus taking shape in the womb, part by part.

Our second example borders on the grotesque. God's creation of the universe in *La Sepmaine* oddly resembles not just gestation and humoral theory, but also the Aristotelian understanding of digestion. Before the separation of light and darkness, the stuff out of which God made the universe was like undigested matter in a stomach. Du Bartas writes:

> Or ces quatre elemens, ces quatre fils jumeaux,
> Savoir est l'air, le feu, et la terre, et les eaux,
> . . . .
> Soit que de toutes pars, condfondant leurs substances,
> Ils facent un seul corps de deux-fois deux essences:
> Ainsi que dans le creux d'un verre christalin
> Le breuvage achelois se mesle avec le vin:
> Ou comme la viande et la boisson subtile
> Chez nous se vont meslant pour se muer en chile. (II. 47–58)

> (Now th'Elements twin-twins (two Sons, two Daughters)
> To wit, the Fire, the Aire; the Earth, and Waters
> . . . .
> Whether in all, their substance they confound,
> And so but one thing of their foure compound:
> As in a *Venice* Glass, before our eyne,
> We see the water intermix with wine:
> Or, in our stomack, as our drink and food
> Doe mingle, after to convert to blood). (II. 63–74)

The four Aristotelian elements—air, water, earth, and fire—were unseparated, mixed in a chaotic, primordial ooze. They were like chyme, or undigested food mixed with acid in the stomach. The unseparated elements are dangerous—they are a 'chaos mutiné', a mutinous chaos that negates Aristotle's theorems of nature. Aristotle assigned each element its own proper place in the cosmos, with its own proper behaviours: air rises to meet air, water falls toward the centre of the earth, and so on. The 'chaos mutiné' is a terrifying breakdown of Aristotelian laws.

So, in Du Bartas's version of the biblical creation story, the universe began as a foreboding, intestinal chyme. Then, God opened his mouth and finished the digestion process. The action of opening his mouth separated all of the Aristotelian elements into their rightful places, converting them from a chaotic mass into a neat organization of elements of which Aristotle would approve. This mimics Aristotle's understanding of digestion, in which the stomach, by the process of *pepsis,* separates liquefied food into its constituent parts.[33] Du Bartas writes:

> Il comprend qu'aussi tost que la bouche de Dieu
> S'ouvre pour assigner à chaque corps son lieu,
> Le feu contre le feu, l'eau contre l'eau se serre,
> L'air se va joindre à l'air, et la terre à la terre. (II. 271–274)
>
> (when the Mouth *Divine*
> Op'ned (to each his proper Place t'assigne)
> Fire flew to Fire, Water to Water slid,
> Aire clung to Aire, and Earth with Earth abid). (II. 295–298)

Note that Du Bartas does not write 'la parole de Dieu', 'the word of God', or something more conventional, but rather, 'la bouche de Dieu', 'the mouth of God'. It is the organ itself that counts. Earlier in the text, Du Bartas had also used an oddly graphic allusion to God's mouth:

> Ainsi le Tout-puissant, avant que, sage, il touche
> A l'ornement du monde, il jette de sa bouche
> Je ne sçay quel beau mot, qui rassemble en un tas
> Tout ce qu'ores le Ciel clost de ses larges bras. (I. 215–218)
>
> (So God, before This Frame he fashioned,
> I wote not what great *Word* he uttered
> From's sacred mouth; which summon'd in a Masse
> Whatsoever now the Heav'ns wide arms embrace). (I. 249–252)

The fact that God does not just speak, but throws from his mouth (*jette de sa bouche*) his command to the world is an unfortunate reminder of what

33 Boylan, 'The Digestive and "Circulatory" Systems in Aristotle's Biology', p. 94.

else may be thrown from one's mouth. Du Bartas's use of graphic physical processes to describe the functioning of the cosmos points to what is most intimate about the body. Intimacy and vastness, according to Morton's analysis, are two sides of the same coin. Morton writes in his analysis of Raphael's speech to Adam in *Paradise Lost*: 'It's not only a vast Universe that Raphael is revealing but also an intimate one—the stars are peopled. This is an amazing affront to the idea of the uniqueness of "mankind", and Raphael prohibits it even as he permits it'.[34] Here, Morton briefly acknowledges the flip side of the sort of uber-cosmic thinking that he lifts up *Paradise Lost* to invoke: the view of the universe taken in this passage is 'intimate' as well as 'vast'. This intimacy relates to what he elaborates on later in *The Ecological Thought*: the uncanniness of the 'mesh', how the hypothetical, lunar Adam and Raphael are 'strange strangers' to the characters speaking, similar yet something other.

We can see in Du Bartas's typically Renaissance treatment of humoral theory and correspondences an echo of what Morton writes about DNA. In the *Ecological Thought,* Morton includes a passage that he likes to iterate throughout his work: 'There is no chimp-flavoured, no human-flavoured DNA; we share 98 percent of our DNA with chimps and 35 percent with daffodils'.[35] For the ecological thought, everything is DNA. For Du Bartas, everything is elements. The same elements that make up stars and trees and cuttlefish make up the human body. Thus, in the sixteenth-century view, the workings of the cosmos are comparable with the workings of the human body. The underlying philosophy behind the microcosm/macrocosm trope resembles what Morton is trying to rekindle with the ecological thought—the idea that all beings, living and non-living, have a base language in common.

Like DNA, Du Bartas's treatment of the human body, like Rabelais's, degrades the whole, fragments it, into those parts that link it to the outside world. It is a great leveller, not just of social classes, but also between animal and human. At once exploiting the most intimate aspects of corporeal existence, it links us to all other bodies. Sixteenth-century ecology does not go to the level of DNA to connect all life forms—it does not need to. Correspondences, humoral theory, macrocosm/microcosm: even if we know today that these have no scientific basis, they still lead to what we can recognize as the ecological thought.

34 Morton, *The Ecological Thought,* p. 22.
35 Ibid., p. 66.

Conclusion

Going back to Morton's analysis of Milton, we can see how Du Bartas's route to the ecological thought differs from his successor's. Whereas Milton's characters arrive at the ecological thought by looking to other planets, Du Bartas does this by looking into the human body, in which he finds a microcosm of the universe. We arrive at the mesh by going deep inside, rather than far outside, the observing body. Influenced by Lucretius, for whom sensory information physically acts upon the observer, Du Bartas expresses this bodily turn with a graphic and medically-informed use of the microcosm/macrocosm trope. In his quest to educate the reader on the glory of the Creator, Du Bartas takes his reader on a journey through the body and through the cosmos, often conflating the two. We delve so deeply into the body of the subject, the human observing his world that, as a consequence, that body dissolves into the stuff of the cosmos. Whereas Milton's text zooms out, out, out, Du Bartas's text zooms in, in, in.

Works Cited

Bakker, Boudewijn, and Diane Webb, *Landscape and Religion from Van Eyck to Rembrandt* (Burlington: Ashgate, 2012).

Bamforth, Stephen, 'Du Bartas et le Merveilleux', in *La Sepmaine de Du Bartas, ses lecteurs et la science du temps: en hommage à Yvonne Bellenger: actes du colloque international d'Orléans, 12–13 juin 2014*, ed. by Denis Bjaï (Geneva: Librairie Droz, 2015), pp. 29–44.

Banderier, Gilles, 'Un "Heureux Phénix"? Renaissance et mort de l'hexaméron (1578–1615)', *Neuphilologische Mitteilungen* 102.3 (2001), 251–267.

Banks, Kathryn, *Cosmos and Image in the Renaissance: French Love Lyric and Natural-Philosophical Poetry* (London: Legenda, 2008).

Bellenger, Yvonne, *Du Bartas et ses divines Semaines* (Paris: SEDES, 1993).

———, 'L'intelligence des animaux: Montaigne et Du Bartas lecteurs de Plutarque', *Revue d'histoire littéraire de la France* 80.4 (1980), 523–539.

Boulton, Elizabeth, 'Climate Change as a "Hyperobject": A Critical Review of Timothy Morton's Reframing Narrative', *Wiley Interdisciplinary Reviews: Climate Change* 7.5 (2016), 772–785.

Boylan, Michael, 'The Digestive and "Circulatory" Systems in Aristotle's Biology', *Journal of the History of Biology* 15.1 (1982), 89–118.

Coupe, Laurence ,'The Ecological Thought', *The Times Higher Education Supplement* 1962 (2010), p. 52.

de Salluste du Bartas, Guillaume, and Yvonne Bellenger, *La sepmaine: (texte de 1581)* (Paris: Nizet, 1981).

Ford, Phillip, 'Lucretius in Early Modern France', in *The Cambridge Companion to Lucretius,* ed. by Stuart Gillespie and Phillip Hardie (Cambridge: Cambridge University Press, 2007), pp. 227–241.

Giacomotto-Charra, Violaine, *La forme des choses: poésie et savoirs dans la sepmaine de Du Bartas* (Toulouse: Presses universitaires du Mirail, 2009).

Hardie, Philip, 'The Presence of Lucretius in Paradise Lost', *Milton Quarterly* 29 (1995), 13–25.

Kany-Turpin, José, 'Une Réinvention de Lucrèce par Guillaume du Bartas', in *La littérature et ses avatars: Discrédits, déformations et réhabilitations dans l'histoire de la littérature: Actes des cinquièmes journées rémoises, 23–27 Novembre 1989*, ed. by Yvonne Bellenger, (Paris: Klincksieck, 1991), pp. 31–39.

King James Bible (1611) (Cambridge: Chadwick-Healy, 1996).

Lamacz, Stéphane, 'La Construction du savoir et la réécriture du *De Rerum Natura* dans *La Sepmaine* de Du Bartas', *Bibliothèque d'humanisme et renaissance* 64.3 (2002), 617–638.

Lestringant, Frank, *L'Atelier du cosmographe* (Paris: Albin Michel, 1991).

Masse, Vincent, 'Newness and discovery in Early Modern France', in *The Invention of Discovery 1500–1700*, ed. by James Dougal Fleming (Surrey: Ashgate, 2011), pp. 167–179.

Morton, Timothy, *Ecology without Nature: Rethinking Environmental Aesthetics* (Cambridge, Mass: Harvard University Press, 2007).

———, *The Ecological Thought* (Cambridge, Mass: Harvard University Press, 2010).

———, 'Ecology as Text, Text as Ecology', *Oxford Literary Review* 32.1 (2010), 1–17.

Naïs, Hélène, *Les Animaux dans la poésie française de la Renaissance* (Paris: Didier, 1961).

Norbrook, David, 'Milton, Lucy Hutchinson, and the Lucretian Sublime', *Tate Papers* 13 (2010), Accessed 18 February 2017. http://www.tate.org.uk/research/publications/tate-papers/13/milton-lucy-hutchinson-and-the-lucretian-sublime. Accessed 27 September 2019.

Reichenberger, Kurt, 'Das Schöpfungsepos des Du Bartas und die Tradition der Antiken Hexamera', *Zeitschrift Für Romanische Philologie* 79.1/2 (1963), 173–186.

Schmidt, Albert-Marie, *La poésie scientifique en France au seizième siècle: Ronsard, Maurice Scève, Baïf, Belleau, Du Bartas, Agrippa d'Aubigné* (Paris: A. Michel, 1938).

Schrijvers, P.H., *Lucrèce et les sciences de la vie* (Boston: Brill, 1999).

Smith, Mick, 'Dark Ecology', *Environmental Politics* 20.1 (2011), 133–138.

Sylvester, Josuah, *Du Bartas: His Divine Weekes and Workes with a Compleate Collection of all the other most delight-full Workes: Translated and written by the famous Philomusus: Josuah Sylvester* (London: Humfray Lownes, 1621).

Taylor, George Coffin, *Milton's Use of Du Bartas* (Cambridge, MA: Harvard University Press, 1934).

Tillyard, E.M.W., *The Elizabethan World Picture* (New York: The Macmillan Company, 1944).
Usher, Phillip John, 'The Holy Lands in Early Modern Literature: Negotiations of Christian Geography and Textual Space', Ph.D. Dissertation. Harvard University, 2004.
———, 'The Revenge of the Mines: Earth-from-Nowhere versus Surfaces-with-Depths' in *Early Modern Visions of Space*, ed. by Jeremie Korta and Dorothea Heitsch (Chapel Hill: UNC Press, forthcoming).
Williams, Arnold, 'Commentaries on Genesis as a Basis for Hexaemeral Material in the Literature of the Late Renaissance', *Studies in Philology* 34.2 (1937), 191–208.
Wilson, Dudley, *French Renaissance Scientific Poetry* (University of London: The Athlone Press, 1974).

About the Author

Stephanie Shiflett holds a PhD in Romance Studies from Boston University. Her dissertation investigates how metaphors that crossed cartography with anatomy expressed subtle, often subversive, theo-political views in many early modern French texts. She presented her work on the cartographer Abraham Ortelius at the Renaissance Society of America Conference and at the University of Toronto in 2017. At Dumbarton Oaks, she developed complementary work on gardens in the scientific poetry of Guillaume du Bartas. A former nineteenth-century specialist, she has also written and presented extensively on George Sand and Honoré de Balzac. She maintains a research blog at mapsandmuscle.wordpress.com.

3. 'When is a meadow not a meadow?': Dark Ecology and Fields of Conflict in French Renaissance Poetry

Jennifer Oliver

Abstract

In poetic responses to the French civil wars, the wounded political body of France is aligned with the ravaged body of the physical landscape in an array of arresting ecological images. By tracing a web of profoundly imbricated commonplaces and analogies concerning fields, bodies, and entrails in particular, this chapter investigates the ways in which the verse of Pierre de Ronsard and Agrippa d'Aubigné both rehearses and decries the unnatural twists and turns of that 'intestine' conflict. Both poets revive ancient expressions of ecological anxiety that disrupt what Timothy Morton has termed 'agrilogistic thought'; but I argue that in their distinctive and sometimes challenging styles, their verse presents (and through syntactic violence, uncannily performs) a still more radical vision of human enmeshment in nature.

Keywords: Ronsard, d'Aubigné, civil wars, intestines, analogy

'I've been kicked in the biosphere'

Political and environmental discourse of the French Renaissance is triangulated through the often violent or medicalized bodily imagery used to describe both state and landscape.[1] In Pierre de Ronsard's poetic reworkings

1 I would like to thank Kathryn Banks for her comments on a very early (not to say embryonic) version of the readings of Renaissance poetry in this piece, and for the generous advice and encouragement she offered for its development at that stage.

Goul, P. and P.J. Usher (eds.), *Early Modern Écologies. Beyond English Ecocriticism.*
Taylor & Francis Group, 2020
DOI 10.5117/9789462985971_CH03

of Age of Gold lore, foundational moments of agricultural 'sin'[2] are brought into dialogue with the religious polemic surrounding France's civil wars; later in the sixteenth century, the 'body' of the French landscape, like the body politic, suffers the ravages of the continuing wars, as lamented by Agrippa d'Aubigné. Images of entrails return insistently; employed, twisting and turning, both to figure France's internal turmoil (as vipers erupting fatally from their mother's belly) and as the site of affect in the face of civil violence. These visceral preoccupations are the strange and anxious ancestors of the tragicomic refrain in *Dark Ecology*, in which Timothy Morton argues that at the core of human-inflicted ecological destruction lies a toxic pattern of 'agrilogistic' thought: 'I've been kicked in the biosphere'.[3] Morton summarizes what he calls the 'agrilogistic algorithm' underpinning religion—and with it the logic of 'civilization'—as consisting of the following 'subroutines': 'eliminate contradiction and anomaly, establish boundaries between the human and the nonhuman, maximize existence over and above any quality of existing'.[4] A little later, he elaborates: 'Agrilogistic space is a war against the accidental. Weeds and pests are nasty accidents to minimize or eliminate'.[5] The broader intellectual tendency that results from this 'algorithm' is the 'cut-along-the-dotted-line' thinking of discrete, easily separated concepts.[6] Crucially, as I will argue, it is possible to attend to non- and differently-agrilogistic voices of the sixteenth century. As I show below, in *De re metallica* (1556) Georgius Agricola defends mining against a plethora of Classical poetic and philosophical invectives, arguing (by means of a contrast with fishing) that it *is* 'natural' for man to plunder the 'bowels of the earth'.[7] Agricola inscribes the nature–culture divide in a slightly different place from many other writers, but he is far from alone in exploring, and worrying about, the limit between the human and the nonhuman. In examining the shifting, messy, and sticky web of metaphors used on both 'sides' of the French conflict, this chapter aims to demonstrate the aptitude of Renaissance analogy for richly illustrating Morton's principle of ecological resistance to 'impossibly tidy boundaries': 'there is no single, independent, definable point at which the meadow

2 See Morton, *Dark Ecology: For a Logic of Future Coexistence*, pp. 38–39.

3 Ibid., p. 119.

4 Ibid., p. 46.

5 Ibid., p. 50.

6 Ibid., p. 93.

7 For a detailed study of entrails in relation to early modern mining, see Phillip John Usher, *Exterranean: Extraction in the Humanist Anthropocene*, Chapters 3–4.

stops being a meadow'.[8] As these texts show us, it is not only nature that tends to defy neat categorization, and refuses to 'cleave to the "Law of Noncontradiction"':[9] making vital and explicit what is latent in Morton's choice of imagery, Renaissance poetry offers its own distinctive (and often radical) modes of aesthetic rebellion. Like the 'arche-lithic' *mauvaises herbes* that, contiguous to the *jardin* enclosed on all sides by the walls of the Bibliothèque nationale de France, and apparently against all odds, continue to thrive beneath the grille found underfoot on the rooftop, the sprawling and protean form of Renaissance poetic analogy repudiates the notion of human dominion over nature.[10]

Where does one commonplace end, and another begin? In one of the most vivid depictions of France's deadly warring factions, d'Aubigné famously describes a beleaguered Mother France, her breast torn and bloodied by her fighting sons. If this passage (taken from d'Aubigné's epic civil war poem *Les Tragiques*) has itself become something of a *topos* among scholars of the French Renaissance, it is precisely this kind of 'background scenery', the aesthetic wallpaper that has come to seem 'given' or even banal through familiarity, that arguably most demands and rewards (re)interrogation from an ecocritical angle. Mother France's body seems to suffer more from the conflict than do those of either of her children. Her body is the battlefield. As a result, her usually nourishing milk is spoiled at its source.

> Je veux peindre la France une mère affligée,
> Qui est entre ses bras de deux enfants chargée.
> Le plus fort, orgueilleux, empoigne les deux bouts
> Des tétins nourriciers; puis, à force de coups
> D'ongles, de poings, de pieds, il brise le partage
> Dont nature donnait à son besoin l'usage;
> Ce voleur acharné, cet Esau malheureux
> Fait dégât du doux lait qui doit nourrir les deux,
> Si que, pour arracher à son frère la vie,
> Il méprise la sienne et n'en a plus d'envie.
> Mais son Jacob, pressé d'avoir jeuné meshui,

8 Morton, *Dark Ecology*, p. 72, p. 73.

9 Morton, *Dark Ecology*, p. 65.

10 Morton coins the term 'arche-lithic' to describe 'a primordial relatedness of humans and nonhumans that has never evaporated', ibid., p. 63. This coexistence extends to language and ideas (including, necessarily, nonsense): 'Plants, specters, and hallucinations return more vividly when you try to prune them', p. 91. On the disruptive potential of the vegetal in the early modern period, see Dominique Brancher, *Quand l'esprit vient aux plantes*.

> Ayant dompté longtemps en son cœur son ennui,
> A la fin se défend, et sa juste colère
> Rend à l'autre un combat dont le champ est la mère.[11]
>
> (I want to paint France as a tormented mother, | carrying two children in her arms. | The stronger of the two arrogantly seizes her | two nourishing teats; then by scratching, | punching, and kicking, he tears up the share | provided by nature to meet his needs; | this remorseless thief, this wretched Esau | lays waste to the sweet milk that ought to feed them both, | such that, in wanting to tear his brother's life from him, | he cares nothing for his own, and is ready to die. | But his Jacob, under strain of starvation, | having long kept his anguish inside his heart, | at last defends himself, and his righteous rage | gives battle to the other, on the field that is their mother).

The focus of these lines takes a *blason*-like poetic impulse of fragmentary bodily description to its extreme, and adds a twist: the limbs of sons and mother intertwine, as 'Esau', from within his mother's embrace, uses the parts of his body as weapons against her. The final line here introduces the image that forms the centrepiece of this chapter: the mother's body, with all of its affective potential, stands at once for the political body of France and for the landscape ravaged by the effects of that political body's disintegration.[12]

A profound ambivalence is reflected in the inescapably interconnected mesh of images and bodies in play. In the context of a civil war, not (quite) cutting off his nose to spite his face, 'cet Esau malheureux' nonetheless inevitably inflicts violence on himself, as on his mother, in giving vent to his fraternal rage.[13] By a strange, inverted reciprocity, d'Aubigné (taking the

11 Agrippa d'Aubigné, *Les Tragiques*, Book I: 'Misères', 80, ll. 97–110.

12 For a fuller account of the workings of this maternal body in d'Aubigné and Ronsard (also discussed here, below), see Keller, Chapter 2: 'Mother France and Her Dysfunctional Family: Religious and National Imagery in Ronsard's *Discours* and *Continuation* and in d'Aubigné's *Tragiques*', pp. 41–76.

13 This horrified impulse of violence towards first the twin and then (by proxy) the mother can be read as an (inverted) manifestation of the process by which the subject emerges in Kristeva's account of abjection (*Pouvoirs de l'horreur*, pp. 20–21): 'L'abjecte nous confronte [...] à nos tentatives les plus anciennes de nous démarquer de l'être *maternelle* avant même que d'ex-ister en dehors d'elle grâce à l'autonomie du langage. [...] Repoussant, rejetant; se repoussant, se rejetant. Ab-jectant. [...] Avant d'être *comme*, 'je' ne suis pas, mais *sépare*, *rejette*, *ab-jecte*. L'abjection, un en sens élargi à la diachronie subjective, est une *pré-condition du narcissisme*.' ('The abject confronts us [...] with our earliest attempts to release the hold of [the] *maternal*

part of the starved and oppressed 'Jacob'), defends his response as 'juste', generating an apologetics of the conflict he simultaneously deplores. Some hundred lines later, this maternal-corporeal imagery is overlaid with that of the ship of state, which, scuttled by the dominant Catholic side, suffers the effects of the same self-destructive impulse that characterizes the civil conflict.

> En cela le vainqueur ne demeurant plus fort,
> Que de voir son haineux le premier à la mort
> Qu'il seconde, autochire, aussitôt de la sienne,
> Vainqueur, comme l'on peut vaincre à la Cadméenne.[14]

> (In this the victor only wins out long enough | to see the object of his hatred meet with death, | which he then follows immediately, self-destroying, with his own, | a victor, but of a Cadmean victory).

Through the viscerally potent coinage 'autochire', d'Aubigné establishes an image of inescapable political-theological enmeshment that finds an echo in the related biological-ecological 'feedback loops' described by Morton in *Dark Ecology*.[15] Evoking the myth of Cadmus, related in Ovid's *Metamorphoses*, in which the soldiers that sprouted from the serpent's teeth sown by Cadmus set to fighting among themselves in 'civil strife' ('iuuentus | sanguineam tepido plangebant pectore matrem'; 'That prime of youth [...] lay writhing

entity even before ex-isting outside of her, thanks to the autonomy of language. [...] Repelling, rejecting; repelling itself, rejecting itself. Ab-jecting. [...] Even before being *like*, "I" am not but do *separate, reject, ab-ject*. Abjection, with a meaning broadened to take in subjective diachrony, is a *precondition of narcissism*'. *Powers of Horror*, p. 13). It is through these attempted self-defining, distancing gestures of 'ex-istence', against the not-quite-self of the twin (and mother-landscape), that d'Aubigné's 'vainqueur' ('victor') comes to discover, or at least make apparent to the poet and reader, his fundamental connectedness to his 'environment'.

14 'Misères', 82, ll. 187–190.

15 See, for instance *Dark Ecology*, p. 7: 'There are *positive feedback loops* that escalate the potency of the system in which they are operating. Antiobiotics versus bacteria. Farmers versus soil, creating the Dust Bowl in the Midwestern United States in the 1930s. Such loops are common in human "command and control" approaches to environmental management, and they result in damage to ecosystems.' The contemporary political fallout, in Syria and elsewhere, of the desertification feedback loop is one of the examples of the disproportionate and racialized impact of climate change on the world's poorest detailed by Naomi Klein in her lecture 'Let Them Drown: The Violence of Othering in a Warming World'. On the importance of recalibrating (i.e. racializing) the study of the (non)human, see Alexander G. Weheliye, *Habeas Viscus*. I am indebted to Sneha Krishnan for reading this piece and giving incisive and expert recommendations for broadening my nonhuman scholarly horizons.

on | Their mother's bloodstained bosom'),[16] d'Aubigné articulates for a second time a particular anxiety concerning fertility gone wrong. As the poet's muse, Melpomène, cries,

> O France désolée! ô terre sanguinaire! [...]
> Sur ton pis blanchissant ta race se débat,
> Là le fruit de ton flanc fait le champ du combat.[17]
>
> (O desolate France! o bloodied land! [...] | Your brood are locked in strife over your blanching bosom, | there the fruit of your belly makes of it the field of battle.)

France's warring factions, whether polluting Mother France's natural bounty or springing nightmarishly from the teeth-seeds of a former monstrous conflict, are both engaged in, or are products of, a twisted agricultural process. D'Aubigné's extraordinary, weirdly compacted syntax here—'de ton flanc' being made to do the double work of describing the origins of 'le fruit' and the setting of the 'combat'—highlights a disturbing point: the 'fields' of France have born a treacherous harvest. What is interior to the nation, and 'natural' to the landscape, can no longer be trusted; emerging from the inside, it lays waste to its own surface.[18] Through his radical manipulation and distortion of language, d'Aubigné fashions a kind of ecological, or 'arche-lithic', poetics.

The proliferation in Renaissance texts of images that challenge internal/external (and indeed human/nonhuman) bodily boundaries is, of course, the subject of copious analysis, such as that by Mikhail Bakhtin, in *Rabelais*

16 Ovid, *Metamorphoses*, ed. by Tarrant, III, pp. 122–124, trans. by Melville.

17 'Misères', 79, ll. 89–96.

18 Kristeva (*Pouvoirs de l'horreur*, p. 84) quotes the anthropologist Mary Douglas on the boundary threat posed by bodily fluids: 'La matière issue de ces orifices (du corps) est de toute évidence marginale. Crachat, sang, urine, excréments, larmes, dépassent les limites du corps [...]. L'erreur serait de considérer les confins du corps comme différents des autres marges'. ('Matter issuing from [the orifices of the body] is marginal stuff of the most obvious kind. Spittle, blood, milk, urine, faeces or tears [...] have traversed the boundary of the body [...]. The mistake is to treat bodily margins in isolation from all other margins'. *Purity and Danger*, p. 122; quoted by Kristeva in French translation: *De la souillure*, p. 137). To read d'Aubigné with Kristeva is to observe that the poet's ingestion-anxiety is linked to a broader ecological one (p. 90): 'Une nourriture ne devient abjecte que d'être un bord entre deux entités ou territoires distincts. Frontière entre la nature et la culture, entre l'humain et le non-humain' ('Food becomes abject only if it is a border between two distinct entities or territories. A boundary between nature and culture, between the human and the nonhuman'. *Powers of Horror*, p. 75).

and His World.[19] But in the specific, weaponized and environmentalized context of the wars of religion, it seems that, for both sixteenth-century poets and twenty-first-century readers and eco-critics, there is even more at stake in asserting the porosity of these boundaries than has previously been explored. At such moments, the concerns of both groups resonate in uncanny ways, as when Morton articulates the same kind of untidy, inescapable, mother-guts-environment association seen above in d'Aubigné's verse:

> One's mother's body *is* the biosphere. And my stomach that feels like it gets kicked really violently with news of extinction isn't my stomach. I'm not talking about little me, the appearance, suffering here. My stomach is also this biosphere. It implies all the not-me beings.
>
> I've been kicked in the biosphere.[20]

French writers and thinkers of the sixteenth century deal in these kinds of 'uncanny', 'weird', or 'loopy' (as Morton calls them) logics as a matter of course.[21] In d'Aubigné's terms: the brother I'm destroying is not only my literal family, or even the political entity 'France' (and what more agrilogistic concept could there be than a nation state?), but the whole, living landscape that holds us, too. The messily enmeshed images in play in the poetic description here pose a challenge to certain rather overly tidy conceptions of Renaissance analogy, such as that famously, if provisionally, evoked by Foucault in 'Le prose du monde':

> Jusqu'à la fin du XVIe siècle, la ressemblance a joué un rôle bâtisseur dans le savoir de la culture occidentale. C'est elle qui a conduit pour une grande part l'exégèse et l'interprétation des textes; c'est elle qui a organisé le jeu des symboles, permis la connaissance des choses visibles et invisibles, guidé l'art de les représenter. Le monde s'enroulait sur elle-même: la terre répétait le ciel, les visages se mirant dans les étoiles, et l'herbe enveloppant dans ses tiges les secrets qui servaient à l'homme.[22]

19 See Bakhtin, *Rabelais and His World*, pp. 162–164 and 221–226; especially p. 226: 'These images create with great artistry an extremely dense atmosphere of the body as a whole in which all the dividing lines between man and beast, between the consuming and consumed bowels are intentionally erased'.

20 Morton, *Dark Ecology*, p. 119.

21 On the 'weird' and the 'loopy', see especially ibid., pp. 6–9.

22 Foucault, *Les Mots et les choses: Une archéologie des sciences humaines*, p. 32.

> (Up to the end of the sixteenth century, resemblance played a constructive role in the knowledge of Western culture. It was resemblance that largely guided exegesis and the interpretation of texts; it was resemblance that organized the play of symbols, made possible knowledge of things visible and invisible, and controlled the art of representing them. The universe was folded in upon itself: the earth echoing the sky, faces seeing themselves reflected in the stars, and plants holding within their stems the secrets that were of use to man).[23]

The 'Renaissance episteme' described by Foucault often proves rather too rigid to describe the wealth and variety of expression across literary genres and forms in the period, and indeed has been deftly dismantled by Ian Maclean: '[Foucault's] quasi-Kantian insistence on conditions of possibility, on formal, preconceptual constraints, and on the limiting factor of the imagination may make him blind to endoxical knowledge, with its untidy edges and imprecisions'.[24] Of course, this challenge to neat frameworks is echoed by those posed in the past few decades of eco-critical thought, following calls by the likes of Bruno Latour, who in 1991 read '[l]es affaires embrouillées' ('mixed-up affairs') of a daily newspaper as giving the lie to conceptions of disciplinary divisions or 'compartiments' ('compartments'): 'Toute la culture et toute la nature s'y trouvent rebrassées chaque jour' ('All of culture and all of nature get churned up again every day') .[25] If the contemporary 'nonhuman turn' asks, 'What if Whitehead, instead of Heidegger, had set the agenda for postmodern thought?',[26] one of the answers to the question posed by Louisa Mackenzie ('What can early modern French literature do for ecocriticism?'), as discussed in the introduction to this volume, is perhaps to offer, in however limited a way, access to pre-Heideggerian, and indeed pre-Kantian, pre-Cartesian, possibilities for thought.

Pursuing this line of thought, this chapter thus traces the interactions between human, political, and environmental bodies in the poetry of Ronsard and d'Aubigné, and asks how the overlapping and intertwining of corporeal analogies for the physical and political landscape might be seen

23 Foucault, *The Order of Things: An Archaeology of the Human Sciences*, p. 19.

24 See Maclean, 'Foucault's Renaissance Episteme Reassessed: An Aristotelian Counterblast', p. 165; by the same author see also *Le Monde et les hommes selon les médecins de la Renaissance*, Chapter 5, 'Postface post-foucaldienne', pp. 111–121.

25 Latour, Bruno, *Nous n'avons jamais été modernes: Essai d'anthopologie symmétrique*, 9. Translation by Catherine Porter: *We Have Never Been Modern*, p. 2.

26 Steven Shaviro, *Without Criteria: Kant, Whitehead, Deleuze, and Aesthetics*, 'Preface: A Philosophical Fantasy', pp. ix-xvi.

to perpetuate—or resist—the 'feedback loop' identified by Morton in the 'virus' of agrilogistic thought.[27] But first, it will be instructive to consider a particular example of the medicalized bodily analogy used by a major actor in the political discourse of the period, not least because in the story of anxiety about poetic, political, and environmental inheritance that unfolds below, the politician and poet in question stands (for Ronsard at least) for a certain kind of father figure.

'Le corps de [n]ostre estat, pasle, maigre et deffiguré'

At the time of France's civil wars it was a commonplace to describe the conflicts as 'intestins'. As John O'Brien has shown, Michel de Montaigne returns several times to this image of internal disorder, made all the more striking and intimate by the fact that he also uses the term to describe the pain he experiences as a sufferer of gallstones.[28] This visceral imagery forms part of the broader analogy of the 'body politic', which in these times of turmoil was described as suffering either an infection or an imbalance of humours, and as consequently being in need of medication, purgation, or amputation, depending on the 'doctor' whose advice was sought. One such self-styled diagnostician was Michel de L'Hospital, the *chancelier de France* (chancellor of France) who, in his political discourses, showed himself to be a mediator figure, prescribing purgation or 'amputation' not of Protestants, nor of Catholic *ligueurs* (leaguers) but, diplomatically enough, of the unnamed 'séditieux' (insurgents) whom he identified as the root cause of the ulcer:

> Il y a beaucoup de choses qui sont en apparence dures et aigres, qui sont neantmoins salutaires [...]: Par mesme façon les meilleures et plus saines medecines sont les plus ameres [...]. Car si nous sommes tous comme un corps, duquel le roy est le chef: il est beaucoup meilleur coupper le membre pourri, que permettre qu'il gaste et corrompe les autres et leur face souffrir mort. S'il y avoit un homme pestiferé, ou infect de lepre, vous le chasseriez de vostre ville: Il y a plus grand'raison de chasser les seditieux.[29]

27 Morton, *Dark Ecology*, p. 7.

28 O'Brien, 'Intestinal disorders'. On visceral and other confessionally charged corporeal imagery in the Wars of Religion, see also Banks, 'Interpretations of the Body Politic and of Natural Bodies in Late Sixteenth-Century France', pp. 205–218; Williams, '"L'Humanité du tout perdue?": Early Modern Monsters, Cannibals and Human Souls', pp. 235–256.

29 L'Hospital, 'Discours du 12 décembre 1560, Parlement, Paris', 403–404, ll. 461–481.

(There are many things that appear to be harsh and sour, and yet they are healthful [...]: In the same way, the best and most wholesome medicines are the most bitter [...]. For if we are all like a body, of which the king is the head, is it much better to cut off the rotten limb, than to let it spoil and corrupt the others and cause them to die. If there were a man infected with the plague, or leprosy, you would chase him from your town: there is even more reason to chase away the seditious).

As Loris Petris observes, the popularity of this metaphor in the period is matched only by the variety and 'souplesse' ('suppleness') of its usage. Even within the corpus of L'Hospital's *Discours*, it is made to argue for radical 'medical' measures (as above), and then, little over a year later, to counsel against armed civil conflict, 'chose qui est non seulement repugnante au nom de chrestien que nous portons, mais à toute l'humanité' ('something which is repugnant not only to the good name of 'Christian' that we uphold, but to all of humanity'), in a passage that packs an affective punch through its additive amplification:

[D]e quels gens de guerre composerons-nous nostre armee? Tels, que nous cuiderons estre de nostre costé, tant capitaines de soldats, seront peut-estre du parti contraire. Et encores qu'ils soyent de mesme religion que nous, je ne sçay comment l'on les pourroit faire combatre quand ils verroyent de l'autre costé ou leurs peres ou leurs fils ou leurs freres ou leurs femmes ou leur plus proches. Et en oultre, la victoire, de quelque costé qu'elle fust, ne pourroit estre que calamiteuse, estant dommageable tant aux vainqueurs qu'aux vaincus, tout ainsi que si les parties du corps se defaisoyent l'une l'autre.[30]

(With which warriors will we build our army? Those who we believe to be on our side, captains as well as soldiers, might be on the other side. And even if they share our religion, I don't know how they could be made to

30 L'Hospital, 'Discours du 3 janvier 1562, Assemblée politique restreinte, Saint-Germain-en-Laye', 437, ll. 117–125. Cf., for an overtly polemic counterpoint, Dorléans, *Avertissement, Des Catholiques Anglois aux François Catholiques, du danger où ils sont de perdre leur Religion, et d'experimenter, comme en Angleterre, la cruauté des Ministres, s'ils reçoivent à la Couronne un Roy qui soit Hérétique. En ceste derniere edition augmenté*; a pamphlet that employs extended medical analogy to describe the ravages of heresy on the body politic: 'Il nous suffit, pour conclure les estranges accez de vostre maladie, de vous representer le corps de vostre estat, pasle, maigre et deffiguré' ('All I need to do, to resolve the strange fits of your illness, is to exhibit to you the pale, gaunt, and deformed body of your state'), p. 11.

> fight when they saw on the other side their fathers, or their sons, or their brothers, or their wives, or their dearest friends. And besides, victory, on whichever side, could only ever be calamitous, being as damaging to the victors as to the vanquished, just as though the parts of the body were destroying one another).

But there will be no such conciliatory tone in any of the poetic texts treated below. The rest of this chapter will focus on two poets—the Catholic Ronsard, and the Reformist d'Aubigné—who describe the conflicts ravaging France in confrontational bodily terms. As seen in the opening example, the body that figures the contested political and physical space of France is distinctly female, and—not coincidentally—is often elided with that other great symbolic female body: Mother Nature. In particular, the analysis here will be concerned with the significance of intestines (or entrails) as a rich locus of overlapping environmental, political, and affective imagery, through which the works of these poets seem to speak to one another across the decades and across confessional divides.[31] As any scholar of early modern French literature knows, the ecology of poetic practices of inspiration in the period makes a mockery of neat ideological, generic, and aesthetic categorizations.

'Ulcerant par sillons les entrailles encloses'

Ronsard's engagement with the French landscape is, in his earlier works at least, more concerned with the exploitation of natural resources than with political strife; perhaps most strikingly, he took a ferocious stance against the deforestation of the Gâtine, the region of his ancestors.[32] In his odes, Ronsard repeatedly deplores mining as opening up the 'entrailles' of rocks, in much the same way—though to different ends—that Georgius Agricola writes of plundering 'the bowels of the earth'. In Book I of *De re metallica*, Agricola gives voice to the critics of mining whose minds he would hope to change:

31 On the history (and the confessional stakes) of cultural responses to the French landscape, see Jean Viard, *Le Tiers espace*.

32 This is explored compellingly by Mackenzie in *The Poetry of Place: Lyric, Landscape, and Ideology in Renaissance France;* see especially Chapter 5, 'The Poet and the Environment: Naturalizing Conservative Nostalgia', pp. 121–145. For Mackenzie, Ronsard, in his impassioned defence of the Gâtine, is not acting in proto-conservationist fervour, nor speaking 'a discourse of a minority oppressed by the violent exclusivity of nationhood', but constructing an exclusive and privileged ideological space.

> The earth does not conceal and remove from our eyes those things which are useful and necessary to mankind, but on the contrary, like a beneficent and kindly mother she yields in large abundance from her bounty and brings into the light of day the herbs, vegetables, grains and fruits, and the trees. The minerals on the other hand she buries far beneath in the depths of the ground; therefore, they should not be sought. But they are dug out by wicked men who, as the poets say, are the products of the Iron Age'. Ovid censures their audacity in the following lines:
>
>> And not only was the rich soil required to furnish corn and due sustenance, but men even descended into the entrails of the earth, and they dug up riches, those incentives to vice, which the earth had hidden and removed to the Stygian shades. Then destructive iron came forth, and gold, more destructive than iron; then war came forth.[33]

For Agricola, who amasses a wealth of such classical poetic arguments against his own cause as a metallurgist, many of these may be countered by the claim that man's 'natural' element is the earth, rather than the sea: 'Indeed, it is far stranger that man, a terrestrial animal, should search the interior of the sea than the bowels of the earth'.[34] But for Ronsard, the moral framing is quite different; drawing on the same Ovidian and Virgilian imagery of the Age of Gold, he laments the advent of technologies including mining, seafaring, and agriculture.[35]

33 Agricola, *De re metallica*, pp. 6–7; *De re metallica libri XII*, p. 4 : 'Terra non occultat et ab oculis remouet ea quæ hominu[m] generi utilia sunt et necessaria, set ut benefica benigna[que] mater maxima largitate sundit ex sese, et in aspectum lucem[que] profert herbas, legumani, fruges, fructus arboru[m]: at fossilia in profunda penitus abstrudit, eruenda igitur non sunt. Quia vero ipsa eruunt homines scelerati, quos ut poetæ loquuntur, ferrea ista ætas progignit, Ovidius eam audaciuam merito insequitur his versibus.

Nec tantum segetes alimenta[que] bebita diues
Poscebatur humus, sed itum est in viscera terrae,
Quas[que] recondiderat, Stygiis[que] admoverat undis,
Effodiuntur opes, irritamenta malorum.
Iam[que] nocens ferrum ferro[que] nocentius aurum
Prodierat, prodit belum.'

34 *De re metallica*, p. 8: 'cum multo magis alienum sit ab hominis terreni animalis uita maris interiora, quam terræ uiscera scrutari.'

35 The most comprehensive study of this commonplace in Ronsard's works is Armstrong, *Ronsard and the Age of Gold*.

In Ronsard's *Hymne de la Justice*, the 'torments' and 'ulcerations' the plough inflicts on the entrails of the earth are definitively the kinds of outrage against Nature that were not perpetrated in the Age of Gold:

> Dieu fist naistre Justice en l'âge d'or ça bas
> Quand le peuple innocent encor ne vivoit pas
> Comme il fait en peché, et quand le vice encore
> N'avoit passé les bords de la boete à Pandore:
> Quand ces mots *Tien* et *Mien* en usage n'estoyent,
> Et quand les Laboureurs du soc ne tourmentoyent
> Ulcerant par sillons les entrailles encloses
> Des champs qui produisoiyent de leur gré toutes choses,
> Et quand les Mariniers ne pallisoyent encor'
> Sur le dos de Tethys pour amasser de l'or.[36]

(God created Justice down there in the Age of Gold, | when the innocent people did not live | as they do now in sin, and when vice had not yet escaped from Pandora's box: | when the words 'yours' and 'mine' were not yet in use, | and when labourers did not torment | and ulcerate with furrows the enclosed entrails | of the fields, which produced of their own accord all that was needed, | and when mariners did not yet throng | the flanks of Tethys to go forth and gather gold).

The descent into the Age of Iron, which in classical sources is marked by the advent of first agriculture and then warfare, is inscribed in both cases on the corporeally depicted surface and depths of the landscape: both the ploughing of fields and (through mining) the forging of weapons are associated with gouging the tender flesh of the earth.[37]

36 Pierre de Ronsard, *Œuvres complètes*, vol. II, 474, ll. 37–46. Further references to this edition will refer to *OC*.

37 It is worth noting here that, by tracing the alterations made by Ronsard over the course of various editions of his works, we find that in his *Franciade* he once replaced a mother's 'ventre', her womb, with 'entrailles', suggesting that, in the context of this feminized landscape-body, the 'entrails' of the earth are, logically enough, also the uterine locus of fertility. It is also striking that, of all of the illustrations in the anatomy book from which the image on these pages is taken, the backdrop to this diagram of the intestines is the most 'earthy'. The images depicting women's anatomy, on the other hand, are 'posed' interior scenes, often on beds, their bodies reclining in a series of eroticized, passive poses. On the evolution of such illustrations, see Bernard Vouilloux, 'Le dispositif anatomique. De la leçon au traité'.

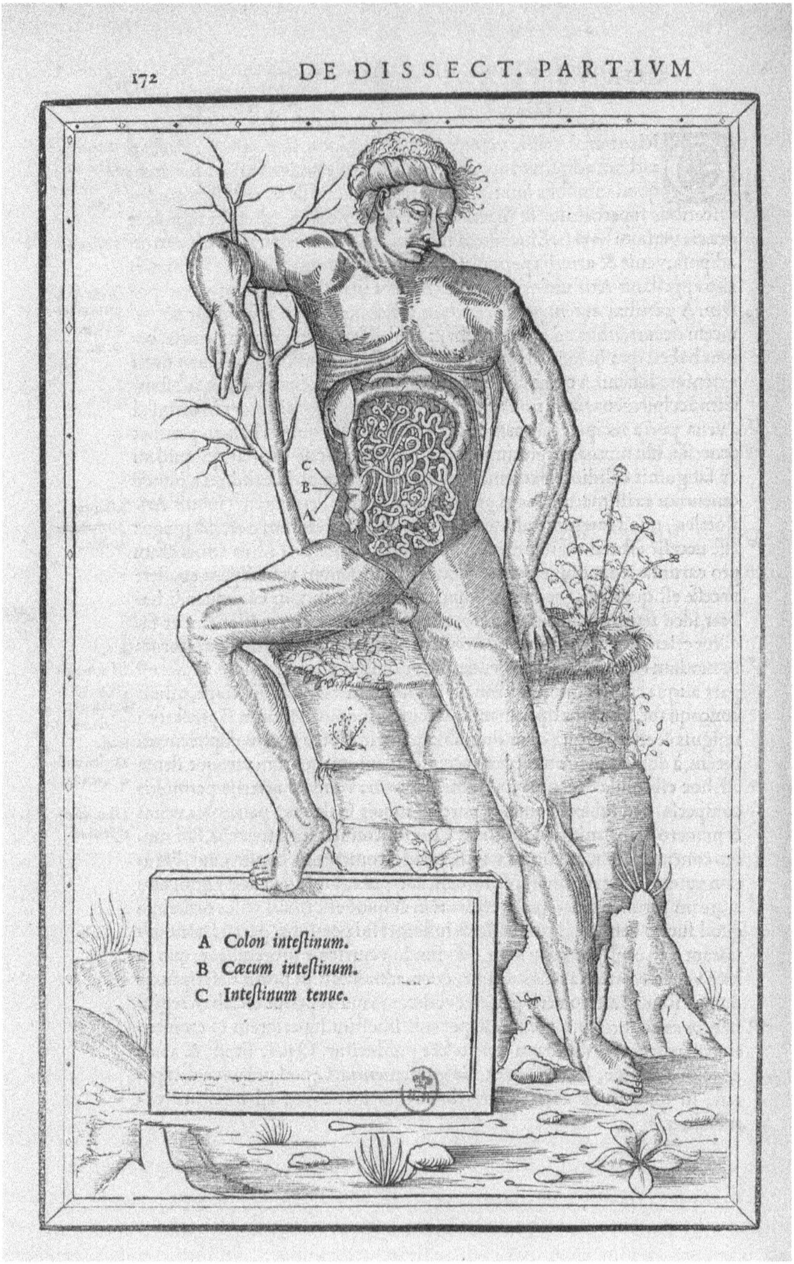

Figure 1: Dissection of the abdomen. Source: Mercure Jollat, in: Charles Estienne, De Dissectione partium corporis (Paris: Simonem Colinaeum, 1545), p. 172. Image from Bibliothèque nationale de France (BnF).

Of course, Ronsard's ideal vision of man living in harmony with Nature is doomed to fail, and Justice, launches a tirade against the 'peuple avorton' ('stunted people') who have ravaged and abused her bounty, and will be reduced to tilling the soil.[38]

Meschant peuple avorton, disoit-elle, est-ce ainsi
Qu'à moy fille de Dieu tu rens un grand merci
De t'avoir si long temps couvé dessous mes ailes,
Te nourrissant du laict de mes propres mammelles?
Je m'en-vole de terre, et je te dis adieu,
Adieu peuple avorton, je t'asseure que Dieu
Vangera mon depart d'un horrible tempeste,
Que ja desja son bras eslance sur ta teste.
Las! où tu soulois vivre en repos planteureux,
Tu vivras desormais en travail malheureux:
Il faudra que tes bœufs aux champs tu aiguillonnes,
Et que du soc aigu la terre tu sillonnes,
Et que soir et matin le labeur de ta main
Nourrisse par sueur ta miserable fain:
Pour la punition de tes fautes malines
Les champs ne produiront que ronces et qu'espines[.][39]

(Vile, stunted people, she said, is this how | you thank me, the daughter of God, | for having kept you so long safely under my wing, | feeding you with the milk from my own breasts? | I am taking flight from the Earth, I bid you farewell, | goodbye runtish race, and I assure you that God | will avenge my departure with a horrific storm, | which his hand is already hurling down on you. | Alas! Where you were accustomed to living in restful abundance, | now you will live in wretched toil: | you will have to spur on your cattle in the fields, | and furrow the earth with sharpened plough, | and day and night your hard labour | will have to feed your miserable hunger: | in punishment for your malign wrongdoings, | the fields will produce only thorns and brambles).

38 Later, d'Aubigné also evokes a female-embodied figure of Justice through his portrayal of the image of Themis in 'La chambre dorée': see *Epic Arts in Renaissance France*, pp. 178–179. If in Ronsard's vision, Justice is linked to an irretrievable past, for d'Aubigné, as Usher argues, this timeless figure also points forward, to 'a future moment of vengeance'.
39 Ronsard, *OC* vol. II, 476, ll. 117–132.

In accordance with Age of Gold mythology[40] the tools of agriculture arrive alongside the first weapons of war, as mankind sinks further into sin and enters the Age of Iron. For Ronsard, then, the horrors of warfare are no more than the natural conclusion of mankind's decline, and so the kind of feminized 'champ de bataille' ('battlefield') represented by d'Aubigné's Mother France is a distant reincarnation of Ronsard's personified—and embodied, in hybrid avian-mammalian form—figure of Justice.[41]

If conflict in general is symptomatic of man's fall from the Age of Gold in Ronsard's poetry, 'intestine' civil wars do seem to hold particular weight. In an invective against the acolytes of 'Predicans' including Théodore de Bèze in the *Continuation du Discours des Misères de ce temps* (1562–1563), France's maternal entrails seem to come alive in an uncanny way: baby vipers, bursting forth fatally from their mother's belly, move the image of 'troubled intestines' on to something altogether more deadly.

> Vous ressemblez encor à ces jeunes viperes,
> Qui ouvrent en naissant le ventre de leurs meres:
> Ainsi en avortant vous avez fait mourir
> La France vostre mere en lieu de la nourrir.[42]

> (What's more, you resemble those young vipers, | who, in being born, rip open their mothers' bellies: | in aborting yourselves in this way, you have killed | your mother France, instead of feeding her).

Later in the same decade, in an epitaph for the Duc de Montmorency (1567/8), an old soldier killed at the Bataille de Saint-Denis, Ronsard's France is no longer a body suffering an internal sickness caused by warring factions, but rather (almost) eviscerates itself, Cato-like, with a weapon of war:

> les François par civiles batailles
> Tournoyent le fer en leurs propres entrailles,
> Espoinçonnez d'infernale fureur [.][43]

> (the French, in civil wars, | twisted the blade in their own entrails, | spurred by infernal fury).

40 Cf. *Metamorphoses* I. 128–147.

41 The notion of Justice leaving the Earth by taking flight, or fleeing, is found in the classical sources: Ovid's *Metamorphoses* (I. 148–149) and Aratus' *Phænomena* (pp. 96–136); but these peculiarly hybrid animal attributes appear to be of Ronsard's own invention.

42 Ronsard, *OC* vol. II, 999, ll. 91–94.

43 Ibid., 161–163.

This last description draws together elements of the other metaphors and analogies discussed so far in an Escher-esque, looping play of images: France holds the sword, a sign (within the symbolic system of Age of Gold mythology) of social, environmental, and political decline, and, in twisting it, tears at the entrails that stand at once for the abundance of Nature and for the internal, factional conflict played out on France's landscape. While for Ronsard the source of blame is distinctly identifiable, his constant recourse to corporeal and ecological analogy speaks of the difficulty in isolating an element so fundamentally imbricated in its surroundings; as this final image shows, Ronsard recognizes what Michel de L'Hospital had identified several years earlier: to attack the internal cause would be to injure the body as a whole.

'Seulement mes entrailles vous ont senti'

Such destruction is all but inevitable for Ronsard's Reformist counterpart, in both the political and the aesthetic realms: in his prefatory notes to the edition quoted here, Frank Lestringant presents d'Aubigné's poetic project in *Les Tragiques* as, among other things, a wrecking of Ronsard's poetic landscape, as well as that of his own juvenalia:

> Du jardin poétique légué par Ronsard et amoureusement cultivé par les émules de la Pléiade, bruissant d'eaux vives et noyé d'ombrages, il ne reste qu'un paysage sinistré, une terre calcinée et à jamais stérile. Animé par une rage iconoclaste, d'Aubigné transporte jusque dans le verger des Muses la guerre civile qui fait rage au-dehors, à travers toute la France des guerres de Religion. C'est avec un bel entrain qu'il saccage les parterres dessinés par lui-même dans *Le Printemps*[.][44]
>
> (Of the poetic garden bequeathed by Ronsard and lovingly cultivated by the disciples of the Pléiade, murmuring with running water and dappled in shade, remains only a desolated landscape, a charred and forever sterile earth. Stirred by an iconoclastic rage, d'Aubigné transports into the orchard of the Muses the civil war that rages on outside, across all of France during the wars of religion. With great enthusiasm, he sets about trashing the flowerbeds that he himself had sketched in *Le Printemps*).

D'Aubigné's apocalyptic vision calls for a return to certain images used by Ronsard to recount the creation narrative: through the ravaged

44 *Tragiques*, p. 8.

body-landscape, Revelation loops back to Genesis.[45] A few lines after the passage that opened this chapter, the development of d'Aubigné's initial affective image sees Mother France decry her sons' destruction and bloodying of her body:

> Elle dit: 'Vous avez, félons, ensanglanté
> Le sein qui vous nourrit et qui vous a porté;
> Or vivez de venin, sanglante géniture,
> Je n'ai plus que du sang pour votre nourriture'.[46]
>
> (She says: 'You have cruelly bloodied | the breast that bore you and feeds you; | now, bloody progeny, live on venom, | for I have only blood left for you to feed on').

In lines that emphasize the conceptual elasticity that 'sein' (incorporating both womb and breast) shares with 'entrailles', the intertwining of limbs noted earlier finds its echo in the mingling of blood with maternal milk: the commixture of bodily fluids that brings these lines into the realm of the abject.[47] This sense of dread at the risk of the surface boundaries and hierarchy of the body politic being dissolved from within by the 'meurtrier de soi-même' ('self-destroyer')[48] unfolds further in the next image—a disease-riddled, putrefying, zombie-like giant, the source of whose dysfunction is plainly located at its core, in the 'ventre':

> Son corps est combattu, à soi-même contraire:
> Le sang pur a le moins, le flegme et la colère
> Rendent le sang non sang; le peuple abat ses lois,
> Tous nobles et tous Rois, sans nobles et sans Rois;
> La masse dégénère en la mélancolie;
> Ce viel corps tout infect plein de la discrasie,

45 For Morton, the darkness of the end is 'seeded' in the beginning: 'agriculture is sin, just like Genesis says [...] something is wrong—*as Genesis had alrady pointed out*', *Dark Ecology*, p. 40. In *Le Contrat naturel*, Michel Serres is also attentive (though with a rather different emphasis) to the parallels between Biblical time and ecological time, and the sense of 'retournement' implied in ecological awareness (for example, see pp. 80, 120).

46 *Tragiques*, I, ''Misères, 80, ll. 127–130.

47 See above, note 17. On the relation of abjected bodies to the landscape underlying the *Tragiques*, and the transformation of Ovidian imagery in this text, see Long, 'Les rivières, sites de massacres et de mémoire dans *Les Tragiques*', pp. 439–454.

48 *Tragiques*, I, ''Misères, 81, l. 134.

[...]
Ce ventre dans lequel tout se tire, tout entre,
Ce faux dispensateur des communs excréments
N'envoie plus aux bords les justes aliments:
Des jambes et des bras les os sont sans moelle,
Il ne va plus en haut pour nourrir la cervelle
Qu'un chime venimeux, dont le cerveau nourri
Prend matière et liqueur d'un champignon pourri.[49]

(Its body is conflicted, pitted against itself: | it has little pure blood, phlegm and choler | make its blood blood no longer; the people overthrow its laws, | all are noblemen and kings, without noblemen, without kings; | the mass degenerates into melancholy; | this old infected body riddled with discrasia, | [...] | This belly into which all is drawn, everything enters, | this treacherous dispenser of the public excrements | no longer sends the rightful nourishment to the peripheries: | the bones of the arms and legs are without marrow, | all that now goes above to feed the brains | is a poisonous vapour, and the brain feeding on this| takes on substance and humour from a rotten mushroom).

If Ronsard's anxiety about France's treacherous progeny earlier found expression in the body horror of vipers erupting from a mother's belly, d'Aubigné pushes the enmeshing of the body politic with the nonhuman still further in this vision of mycotic invasion. More than ever, the chances of political-medical remedy seem remote.

In the initial image of Mother France, the impact of the extended corporeal metaphor is heightened by the woman's specifically maternal vulnerability. This effect is intensified through the multiplying intestinal images, since elsewhere in the poem 'entrailles' are not only shown to have physical significance but are also, metonymically, symbolic of affective response, as in d'Aubigné's account of a son's response to the sight of his dying father:

L'enfant rompt ces propos: 'Seulement mes entrailles
Vous ont senti, dit-il, et les rudes batailles
De la prochaine mort n'ont point épouvanté

49 *Tragiques*, I, 'Misères', 81 ll. 141–146, 150–156. For a medicalized reading of this and other bodies in d'Aubigné, see Losse, *Syphilis: Medicine, Metaphor, and Religious Conflict in Early Modern France,* Chapter 6, 'Tragic Afflictions: D'Aubigné's *Tragiques*', pp. 106–120. See also Prat, *Les Mots du corps: un imaginaire lexical dans les* Tragiques *d'Agrippa d'Aubigné.*

L'esprit instruit de vous, le cœur par vous planté.
Mon amour est ému, l'âme n'est pas émue;
Le sang, non pas le sens, se trouble à votre vue;
Votre blanche vieillesse a tiré de mes yeux
De l'eau, mais mon esprit est un fourneau de feux'[.][50]

(The child interrupts this speech: 'Only my entrails | felt you[r pain], he says, and the pitiless batallions |of imminent death did not trouble | the mind trained by you, the heart steadied by you. | My love is moved, my soul is not; | my blood, but not my sense, is shaken by the sight of you. |Your pale old age drew water from my eyes, but my mind is a flaming furnace').

This scene of spectatorship locates sensibility (but not sense) in the entrails, and models the reaction of the ideal reader who, presented with the graphic imagery splattered over d'Aubigné's canvas, will be moved and yet resolute in their Reformist loyalty. By insisting on the direct, sympathetic resonance between spectacle and tears (as between 'sang' and 'sang' in line 923), to the exclusion of the mind—which might otherwise be assumed to mediate here—d'Aubigné's scene of pathos might also be said to anticipate twenty-first-century theories of affect.[51] We note here, too, a second occurrence of 'entrailles' rhymed with 'batailles' ('battles'), a pairing that recurs frequently in poetry of the civil wars, and perhaps underpins the frequency of the intestinal metaphor.

In the seventh and last book of *Les Tragiques*, anxieties concerning the potential disintegration of the state are once more figured through the emergence of unnatural offspring, though this time the family takes first avian, and then reptilian form:

Le ciel n'est plus si riche à nos nativités,
Il ne nous départ plus de générosités,
Ou bien nous trouverions de ces engeances hautes
Si les mères du siècle y faisaient moins de fautes:
Ces œufs en un nid ponds, et en l'autre couvés,

50 *Tragiques*, IV,'Les Feux', 215 ll. 933–940.

51 See, for example, Massumi's *Parables For the Virtual: Movement, Affect, Sensation*, Chapter 1, 'The Autonomy of Affect', pp. 23–45, on autonomous (galvanic skin) responses to affective (as distinct from emotional) stimuli. Katherine Ibbett offers thought- and feeling-provoking reflections on 'Affect' and the early modern in two recent publications: 'When I Do, I Call it Affect', *Paragraph* 40.2 (2017), 244–253; and *Compassion's Edge: Fellow-Feeling and its Limits in Early Modern France* (Philadelphia: University of Pennsylvania Press, 2017).

Se trouvent œufs d'aspic quand ils sont éprouvés;
Plus tôt ne sont éclos que ces mortels vipères
Fichent l'ingrat fiçon dans le sein des faux pères.[52]

(The heavens no longer look so kindly on the births of our kind, | we no longer inherit moral noblesse, | or rather we might find honorable offspring | if the mothers of our age were more free of sin: | these eggs, laid in one nest and incubated in another, turn out to be asp eggs; | no sooner are they hatched than those deadly vipers | plunge their thankless fangs into the breast of their supposed fathers).

In this strange and rare instance, accusations of infidelity against the wife of prominent Huguenot general Henri de Bourbon provoke a gender switch; it is now a father's breast bloodied by the attack of a monstrous progeny, the venom not feeding but emanating from these illegitimate, usurping vipers in the nest. In the final, apocalyptic section of d'Aubigné's epic work, the 'venin' evoked by Mother France in the opening book silently returns, ouroboros-like, to bite the breast of the would-be heads of a stumbling Reformist lineage.

Conclusion

So you can get stuck in [...] the tragedy of realizing that trying to escape the web of fate is the web of fate. Yet within the melancholia is an unconditional sadness. And within the sadness is beauty. [...] Laughter inside tragedy. Comedy, the possibility space of which tragedy is a rare form. Comedy, the genre of coexistence.[53]

As these examples make abundantly and dizzyingly apparent, Ronsard's and d'Aubigné's figuring of both political and ecological environments through bodies, and entrails in particular, is bound up with the poets' reading and 'digestion' of Age of Gold poetic sources and *topoi*. As we have seen, the symbolism of intestinal imagery is dual, with the combination of physical and affective reference packing an extra punch. But the broader picture that has emerged is not a simple triangulation of political and

52 *Tragiques*, VII, 'Jugement', 310–311, ll. 201–208.
53 Morton, *Dark Ecology*, p. 119.

environmental themes through the shared image of the body. It is something more like a web, or maze, of images, with birds brooding on snakes' eggs, vipers hatching or re-emerging from a mother's belly as uncanny entrails, winged mammals taking flight, and fertility run amok in the 'peuple avorton' decried by Ronsard's Justice. The referential slipperiness of a single term such as 'entrailles' alerts us to the twists and turns to follow as metaphors in French Renaissance poetry are extended or shape-shifted, sometimes leading, infuriatingly, back to where we started. But rather than ending on a pessimistic note, I want to make a positive case for the 'weirdness' of Renaissance analogy.

In the introduction to an earlier eco-critical work, *Ecology Without Nature: Rethinking Environmental Aesthetics*, Morton points to a parallel between the dual fetishization of the 'natural' and the 'feminine': 'Putting something called Nature on a pedestal and admiring it from afar does for the environment what patriarchy does for the figure of Woman. It is a paradoxical act of sadistic admiration'.[54] The 'dark ecology' that Morton proposes in his 2016 work is a continuation of this line of critical thinking, a rebuttal of and remedy to that toxic conception of 'Nature-with-a-capital-N'. Returning to the emblematic first passage quoted in this chapter, we might reconsider, in this light, the curious move by which d'Aubigné shifts from lamenting the laceration of the breast of Mother France to defending the 'just' cause of battle on the 'field' of her body. This might seem, at first, to exemplify a combination of these sadistic impulses. But the weirdly entangled and overlapping images of this poetic landscape seem, on the contrary, to confront and confound the 'impossibly tidy boundaries' presupposed by agrilogistic thought. And as both civil war poets come to acknowledge, from their supposedly distinct 'sides', there is no escape from the interconnectedness of things, from coexistence.

> One's mother's body, the biosphere isn't some abject disgusting thing from which one must distinguish oneself. Underneath the disgust and the horrific uncanny is a type of melancholia, another Freudian term pointing to the indigestible physical and psychic memory trace of other

54 Morton, *Ecology Without Nature: Rethinking Environmental Aesthetics*, p. 5. This kind of observation has, of course, long been made by feminist critics, most notably perhaps in this context by Carolyn Merchant in *The Death of Nature*. For example, Merchant writes of feminist celebrations of Gaia and other goddesses: 'If women overtly identify with nature and both are devalued in modern Western culture, don't such efforts work against women's prospects for their on liberation? [...] Such actions seem to cement existing forms of oppression against both women and nature, rather than liberating either' ('Preface: 1990', in *Death of Nature*, p. xvi).

> beings within oneself. Indigestible, because once you think you've gotten rid of one, along comes another like the heads of the many-headed hydra.[55]

D'Aubigné's fundamentally pessimistic, eschatological perspective engenders, almost despite itself, a dark ecology of sorts. If both he and Ronsard express horror at—and through—scenes of ecological destruction, their twinned, warring responses offer two contrasting directions of attempted escape from what Morton terms the 'agrilogistic feedback loop': backwards, to a time before the advent of 'agricultural sin' (though as Morton argues 'busting out only ever ends up doubling down on what it was trying to escape');[56] or forwards, inexorably down, in order to get out. As Morton puts it, 'Let's make it down into the sadness and proceed further down from there'.[57]

Works Cited

Agricola, Georgius, *De re metallica libri XII* [...] (Basel: Froben, 1556).

———, *De re metallica*, trans. by Herbert Clark Hoover and Lou Henry Hoover (New York: Dover, 1950).

Armstrong, Elizabeth, *Ronsard and the Age of Gold* (Cambridge: Cambridge University Press, 1968).

d'Aubigné, Agrippa, *Les Tragiques*, ed. by Frank Lestringant, 2nd edn (Paris: Gallimard, 2003).

Bakhtin, Mikhail, *Rabelais and His World*, trans. by Hélène Iswolsky (Bloomington, IN: Indiana University Press, 1984).

Banks, Kathryn, 'Interpretations of the Body Politic and of Natural Bodies in Late Sixteenth-Century France', in *Metaphor and Discourse*, ed. by Andreas Musolff and Jörg Zinken (Basingstoke: Palgrave and Macmillan, 2009), pp. 205–218.

Brancher, Dominique, *Quand l'esprit vient aux plantes: Botanique sensible et subversion libertine (XVIe-XVIIe siècles)* (Geneva: Droz, 2015).

Dorléans, Louis, *Avertissement, Des Catholiques Anglois aux François Catholiques, du danger où ils sont de perdre leur Religion, et d'experimenter, comme en Angleterre, la cruauté des Ministres, s'ils reçoivent à la Couronne un Roy qui soit Hérétique. En ceste derniere edition augmenté* (Paris: (n.s.), 1588).

55 Morton, *Dark Ecology*, pp. 118–119.

56 Ibid., p. 127.

57 Ibid., p. 119.

Douglas, Mary, *Purity and Danger: An Analysis of Concepts of Pollution and Taboo* (Abingdon and New York: Routledge, 2003; first published in 1966).

———, *De la souillure. Essai sur les notions de pollution et de tabou*, trans. by Anne Guérin (Paris: Maspéro, 1971).

Foucault, Michel, *Les Mots et les choses: Une archéologie des sciences humaines* (Paris: Gallimard, 1966).

———, *The Order of Things: An Archaeology of the Human Sciences* (London and New York: Routledge Classics, 2002).

Keller, Marcus, *Figurations of France: Literary Nation-Building in Times of Crisis (1550–1650)* (Newark: University of Delaware Press, 2011).

Klein, Naomi, 'Let Them Drown: The Violence of Othering in a Warming World', published in *London Review of Books*, 38.11 (2 June 2016), 11–14.

Kristeva, Julia, *Pouvoirs de l'horreur. Essai sur l'abjection* (Paris: Éditions du Seuil, 1980).

———, *Powers of Horror. An Essay on Abjection*, trans. by Leon S. Roudiez (New York: Columbia University Press, 1982)

Latour, Bruno, *Nous n'avons jamais été modernes: Essai d'anthopologie symmétrique* (Paris: La Découverte, 1991).

———, *We Have Never Been Modern*, trans. by Catherine Porter (Cambridge, MA: Harvard University Press, 1993).

L'Hospital, Michel de, 'Discours du 12 décembre 1560, Parlement, Paris' and 'Discours du 3 janvier 1562, Assemblée politique restreinte, Saint-Germain-en-Laye' in *La Plume et la Tribune: Michel de L'Hospital et ses discours (1559–1562)*, ed. by Loris Petris (Geneva: Droz, 2002).

Long, Kathleen Perry, 'Les rivières, sites de massacres et de mémoire dans *Les Tragiques*', in *Illustrations inconscientes: écritures de la Renaissance – Mélanges offerts à Tom Conley*, ed. by Phillip John Usher and Bernd Renner (Paris: Classiques Garnier, 2014), pp. 439–54.

Losse, Deborah, *Syphilis: Medicine, Metaphor, and Religious Conflict in Early Modern France* (Columbus: Ohio State University Press, 2015).

Mackenzie, Louisa, *The Poetry of Place: Lyric, Landscape, and Ideology in Renaissance France* (Toronto: University of Toronto Press, 2011).

Maclean, Ian, 'Foucault's Renaissance Episteme Reassessed: An Aristotelian Counterblast', *Journal of the History of Ideas*, 59.1 (1998), 149–166.

———, *Le Monde et les hommes selon les médecins de la Renaissance* (Paris: CNRS, 2006).

Massumi, Brian, *Parables For the Virtual: Movement, Affect, Sensation* (Durham, NC: Duke University Press, 2002).

Merchant, Carolyn, *The Death of Nature: Women, Ecology, and the Scientific Revolution* (San Francisco: HarperOne, 1980; issued with new preface 1989).

Morton, Timothy, *Dark Ecology: For a Logic of Future Coexistence* (New York: Columbia University Press, 2016).

———*Ecology Without Nature: Rethinking Environmental Aesthetics* (Cambridge: Harvard University Press, 2009).

O'Brien, John, 'Intestinal disorders', in *Writers in Conflict in Sixteenth-Century France: Essays in honour of Malcolm Quainton*, ed. by Elizabeth Vinestock and David Vine (Durham: School of Modern Languages and Cultures, Durham University, 2008), pp. 239–258.

Ovid, *P.Ovidi Nasonis Metamorphoses*, ed. by Richard J. Tarrant (Oxford: Oxford University Press, 2004).

———*Metamorphoses*, trans. A.D. Melville (Oxford: Oxford University Press, 1986; reissue 2008).

Prat, Marie-Hélène, *Les Mots du corps: un imaginaire lexical dans les 'Tragiques' d'Agrippa d'Aubigné* (Geneva: Droz, 1996).

Ronsard, Pierre de, *Œuvres completes*, ed. by Jean Céard, Daniel Ménager and Michel Simonin, 2 vols. (Paris: Gallimard, 1994).

Serres, Michel, *Le Contrat naturel* (Paris: Le Pommier, 2018; first publ. 1990).

Shaviro, Steven, *Without Criteria: Kant, Whitehead, Deleuze, and Aesthetics* (Cambridge, MA: MIT University Press, 2009).

Usher, Phillip John, *Exterranean: Extraction in the Humanist Anthropocene* (New York: Fordham University Press, 2019).

Viard, Jean, *Le Tiers espace. Essai sur la nature* (Paris: Méridiens Klincksieck, 1990).

Vouilloux, Bernard, 'Le dispositif anatomique. De la leçon au traité,' in *Les Sciences et le livre. Formes des écrits scientifiques des débuts de l'imprimé à l'époque moderne*, ed. by Joëlle Ducos (Paris: Hermann, 2017), pp. 125–141.

Weheliye, Alexander G., *Habeas Viscus: Racializing Assemblages, Biopolitics, and Black Feminist Theories of the Human* (Durham: Duke University Pres, 2014).

Williams, Wes, '"L'Humanité du tout perdue?": Early Modern Monsters, Cannibals and Human Souls', *History and Anthropology* 23.2 (2012), 235–256.

About the Author

Jennifer Oliver is a Departmental Lecturer in French at Worcester College, where her research is centred on sixteenth-century French literature, culture, and thought. Her first book, *Shipwreck in French Renaissance Writing: The Direful Spectacle*, was published by Oxford University Press in August 2019; it sets fictional and allegorical shipwrecks alongside the eyewitness accounts of travel writers in order to explore the relationship between the material and the metaphorical. Her current research project is concerned with how early modern French writers (including Rabelais, Ronsard, Montaigne, and d'Aubigné, but also technical writers including Philibert de l'Orme) contemplated the connections and tensions between poetics, technology, and the natural environment. Her broader research interests include the intersections between literary studies and the history of science, medicine, and technology; art and architectural history; nonhuman and eco-critical theory; and metaphor theory.

4. Equipment for Living with Hyperobjects: Proverbs in Ronsard's *Franciade*[1]

Kat Addis

Abstract
In poetic responses to the French civil wars, the wounded political body of France is aligned with
This chapter calls attention to the proverbs that punctuate Ronsard's unfinished epic poem *La Franciade.* It proposes that, as literary forms, the proverbs share the massively distributed, viscous and non-local qualities of Timothy Morton's hyperobjects. In Ronsard's 1572 epic the significance of the aftermath of the Trojan war turns out to have extended far beyond Virgil's *Aeneid* and the foundation of Rome, to the foundation of Paris and a yet-to-be-realized early modern French empire. *La Franciade*'s proverbs challenge their readers to perceive and respond to these vastly expanded relations, even as they progress through apparently local narrative time., On this basis, they might also equip their readers to engage with the dissonant scales of ongoing global ecological crisis.

Keywords: epic, proverbs, hyperobjects, empire.

In the proverb 'These things are sent to try us', the word *things* expands like Mary Poppins's handbag to contain possibilities of every possible size: a hiccup, a strong gust of wind, a hurricane, global warming. Proverbs such as this one might at first ring hollow because what is being said can change so drastically (mild indigestion or environmental degradation at an unprecedented scale) even as the form remains remain exactly the

1 My title is a nod to Burke's 'Literature as Equipment for Living'.

Goul, P. and P.J. Usher (eds.), *Early Modern Écologies. Beyond English Ecocriticism.* Taylor & Francis Group, 2020
DOI 10.5117/9789462985971_CH04

same: 'These things are sent to try us.' But proverbs are clearly *not* hollow. Proverbs are forms of speech that point towards a content that is, in some fundamental sense, elsewhere (e.g. a shared language or knowledge gathered in other contexts) or, we might say, elsewhen (e.g. knowledge that is outside of chronology, or from another time). Following Craig Dionne, who studies Hamlet's criticism of Osric's stultifying use of commonplace, I see proverbs as 'yeasty'—they are not insignificant, they grow into things that are 'more than the sum of [their] parts'.[2] More specifically in this chapter, I advance the argument that proverbs function within a system of what Timothy Morton calls hyperobjects, that is, things that are 'massively distributed in time and space', 'viscous' and 'nonlocal', and which 'involve profoundly different temporalities than the human ones we are used to'.[3] Global warming is a hyperobject—it exists, but it never exists *in its totality, right here, right now*. It is not an object we can touch or see, but a hyperobject that we can access (but only ever partially) via its effects and via phenomena: via the hand that turns the ignition key of a car, via the temperature in a given place on a given day that melts specific ice. The present chapter therefore asks: what can proverbs, as access points to hyperobjects with which we have been living for a long time now, teach us about being articulate in the Anthropocene? What follows is a discussion of ecological reading—ecological not (for the most part) thematically, but theoretically. I analyze a number of proverbs contained in Pierre de Ronsard's unfinished epic, *La Franciade / The Franciad* (1572), a poem that tells the story (based largely on that of Aeneas) of Francus, the son of Hector and mythological founder of France.[4] If this text can serve us here, it is for two reasons: (1) it contains over thirty proverbs, whose presence is indelibly marked by the use—since the original 1572 edition—of quotation marks that identify them and keep them from being absorbed by their narrative surroundings; and (2) these seemingly mute messengers function, collectively, to point the hero's and the reader's attention towards the hyperobject that is Renaissance epic's imperial teleology writ large (in the hero's fate/destiny, and the rise of *imperium*/empire). As such, Ronsard's proverbs, which have nothing to say directly about climate crisis, offer us reading lessons for our current global ecological crisis nonetheless. They force reckonings with that which we can only perceive in parts and remind us that we should not turn away from 'these things' that 'are sent to try us'.

2 Dionne, *Posthuman Lear*, p. 20.

3 Morton, *Hyperobjects*, p. 1.

4 For a general introduction to the *Franciade*, see Usher, 'La Franciade'.

The Sea Speaks Imperfectly

In the third book of *La Franciade*, a shipwrecked Francus finds himself on the island of Crete, where the sea goddess Leucothea reaches her head out of the water in order to encourage the demoralized hero. To comfort him, she offers up two proverbs that pit patience against urgency.

> Enfant royal qui dois donner naissance
> A tant des rois: '*la seule patience*
> *Rompt la fortune, et mal ne peut s'offrir*
> *Qui ne soit doux quand on le veut souffrir.*'
> Sois courageux: '*Toute rude adventure*
> *Par temps se fait douce quand on l'endure*':
>
> (Royal child who must give birth
> To so many kings. 'Patience alone will
> Tame fortune and there is not toil which
> Is not sweet when suffered voluntarily.'
> Take courage and remember this: 'austere fortune
> Sweetens over time as we endure it').[5]

Leucothea voices these proverbs as a version of the sea—her name means 'goddess of the white foam' (from the Greek *leukos*: white, and *thea*: goddess). She is the foam on the wave tops. In their repetitiveness, moreover, the two proverbs mimic the very motion of the tide. If Leucothea speaks with an ecological voice, we should not be surprised that she expresses her advice in a form that is easy to dismiss.[6] What Leucothea's proverbs *say*—beyond advocating patient endurance—is not immediately clear. It is certainly strange to advise an epic hero to *be patient.* Here, Francus is stuck in Crete—the Carthage of *La Franciade*, populated this time by two amorous women, the sisters Clymène and Hyante.[7] Crete is a dangerous place to end up, of

5 Ronsard, *La Franciade / The Franciad*, 3. 269–265. Italics mine.

6 For a useful survey on ways of attending to the sea as an ecological voice see Yaeger, 'Editor's Column: Sea Trash, Dark Pools, and the Tragedy of the Commons' and Hofmeyr and Bystrom's special issue of *Comparative Literature* on the topic of 'Oceanic Studies'.

7 *La Franciade*'s depiction of Clymene and Hyante's rapacious love for Francus is another cause for proverbs to proliferate. A study would be welcome on the topic of the misogynistic proverb in *La Franciade*, of which there are numerous examples such as: 'Tout cueur de femme est aspre à la vengeance' ('Every woman's heart lusts for vengeance'). Ronsard, *La Franciade / The Franciad*, 1. 170.

course—this is where the Trojans, led by Aeneas, founded their short-lived and mistaken settlement.[8] All signs indicate that what is needed from Francus at this moment is not patience (which could keep him in Crete indefinitely), but decisive action. There is, however, a sense in which Leucothea is perfectly right. This epic hero's destiny is fated—thus, there is nothing that he can do; he must be patient in order to wait for his fate to catch up to him. The proverbs point away from the present moment and towards that destiny.

When Leucothea introduces her patience proverbs by addressing Francus as he 'qui doi[t] donner naissance | a tant des rois' ('who must give birth | To so many kings'), she refers to Jupiter's divine decree. This *must* happen. As such, she directs attention back to promises of empire, as when Virgil's Jupiter promises Venus that her son will be a great foundational figure: 'his ego nec metas rerum nec tempora pono; | imperium sin fine dedi' ('For these I set no bounds in space or time; but have given empire without end').[9] Ronsard's Jupiter makes an identical statement (in a slightly more long-winded version):

> De ce grand Roy je n'ay borne l'empire,
> L'an si dispos qui se change & se vire
> Cassant des Rois les scepters & la loy,
> Ne Perdra point l'empire de ce Roy,
> Qui florira comme une chose ferme
> En son entier, sans limite & sans terme.
>
> (For this king's empire I set no limits.
> The nimble year that changes and turns,
> Which shatters the scepters and reigns of kings,
> Will not put an end to this king's empire,
> Which will grow, solid through and through,
> With neither limit nor end).[10]

Given such a destiny—such a *hyperobject*—what Francus does about it may not much matter. Leucothea's proverbs point Francus in the direction

8 Braybrook, 'The Aesthetics of Fragmentation in Ronsard's *Franciade*' p. 8. Crete is the birthplace of Teucer, ancestor of the Trojan royal line (2. 618–19). On memories of Troy as a dangerously anti-epic force in the *Aeneid* too see Quint, *Epic and Empire*, p. 58. On the role of Crete in the *Franciade* in particular, see Usher, '*Non haec litora suasit Apollo*: la Crète dans la *Franciade* de Ronsard' and 'La Crète épique: La *Franciade* et la tradition des *isolarii*', and *L'Aède et le géographe*, pp. 144–161.

9 Virgil, *Aeneid*, 1. 278–279.

10 Ronsard, *La Franciade / The Franciad*, 1. 265–268.

of this vast empire, this epic imperialist destiny that spans both temporal and spatial expanses. These proverbs then, through their strange call for patience, provide Francus with access to that imperial hyperobject. Without this connection their presence would not be justified. The problem is that inaction is not very heroic, a risk that Leucothea perceives immediately. She quickly gives Francus further instructions—he must bury his friend and pursue Hyante—but this extra information perhaps fails to undo the impression of fundamental stasis that has been introduced by the patience proverbs. One of Leucothea's Virgilian parallels is Venus, who advises Aeneas: 'perge modo et, qua te ducit via, derige gressum' ('Only go forward and where the path leads you, direct your steps!').[11] The contrast in their advice reveals a fundamentally different viewpoint on the respective heroes: Aeneas is an active hero who keeps going; Francus must have patience because the empire he seeks is now massively temporally distended. Shortly after this, Francus describes himself being dragged along by fortune: 'les destins [...] contre mon gré me traisnent' ('the fates [...] drag me on against my will').[12] Francus's inertia, the symptom of a stillness that lurks within heroism throughout *La Franciade*, undermines the paradoxical construct of epic imperial glory, which must be both destined and acquired.[13] This anti-epic stillness is reproduced in the space of the proverbs on the printed page. Leucothea's proverbs point Francus to the hyperobject that is his epic destiny—but they do not tell him exactly *how*, as a human, to react to it.

Several days after his shipwreck, now on Cretan soil, Francus begins his epic lament as follows: 'Heureux trois fois ceux que la bonne Terre | Loing de la vie en long repos enserre' ('Thrice happy those whom, far from life, | The good Earth commits to eternal rest'),[14] echoing of course the similar lament made by Aeneas while the latter is enduring a difficult storm: 'o terque quaterque beati, | quis ante ora partum Troiae sub moenibus altis | contigit oppetere!' ('O thrice and four times blest, whose lot it was to meet death before their fathers' eyes beneath the lofty walls of Troy!').[15] The differences are striking: Aeneas has every reason to believe that he is about to die, so he wishes he could have died at Troy; Francus has, for his part, already defeated a giant, slept safely in a bed, and been welcomed and feasted by his hosts.

11 Virgil, *Aeneid*, 1.401.

12 Ronsard, *La Franciade / The Franciad*, 3. 406–407.

13 See also the hilarious moment of Francus's awakening into heroism in Book 1 when he is compared to a Mars (Ronsard, *La Franciade / The Franciad*, 1. 844–848). Here, again, there is something quite unconvincing about Francus's personal heroism.

14 Ibid., 1. 200–202.

15 Virgil, *Aeneid*, 1. 94–95.

His lament is prompted by the memory of his shipwreck, not by the wish for a more noble death. Even being in a position to have this memory is, one might say, an indulgence on Francus's part, a result of his idle hanging around: 'Luy *se laissant* en larmes consommer | S'alla planter sur le bord de la mer' ('Allowing himself to dissolve into tears | He went to sit on the seashore').[16] When Aeneas lands on the shores of Libya, he immediately sets about surviving; he sleeps, hunts, and eats. Francus consumes himself in tears and goes to brood on the beach. Like a teenager in a sitcom, or like a disgruntled climate sceptic, he does not know how to react to the hyperobject he perceives but cannot grasp.

Francus's lament is prompted by the sight of his wrecked ship, which floats half-submerged in the bay, 'couvert de falaize et de bourbe' ('covered in sand and slimy mud').[17] But it is *not just* a wrecked ship: it is a reminder of elsewhere and elsewhen; it has a different temporality and set of references from its surroundings. It is there according to hyperobjective logic, connecting to another aspect of Francus's epic destiny: namely, via the Virgilian intertext, Troy. Aeneas's version of the lament explicitly references the walls of Troy as a better place to have died. Its absence in Francus's reprise (the place to die has become 'la bonne terre', 'the good land') indicates not that it has been forgotten, but that the role of Troy in the imagination of the epic hero has changed. As Hector's son, Francus is part of the next generation, who did not define themselves in that original war but who deal with its aftershocks—just as the hyperobject of global warming is to be faced by generations not directly responsible for it. Francus stands not only between generations but also between texts, because he is, of course, the result of Ronsard reading Virgil. The wrecked ship then, as a kind of physical proverb, reminds Francus about Troy but does not offer it to him (or us) as a graspable reality because the Troy he 'remembers' is textual—and as part of the massive epic hyperobject of imperial destiny, it is *always* out of place and time. It is, indeed, in the nature of hyperobjects to collapse temporality. The ship's connection to Troy (via Aeneas's shipwreck lament) thus suspends Francus anachronistically between Troy and Paris: 'Voirray-je point une Troyenne plaine, | Voirray-je point ceste gauloise Seine' ('Will I ever see a Trojan field, | Will I ever see the Gallic River Seine').[18] It is clear at this moment that Francus is participating in a hyperobjective network. The proverbs that Leucothea offers to Francus provide instruction that is relational and, in our sense, ecological. It is not

16 Ronsard, *La Franciade / The Franciad*, 3. 191–192.

17 Ibid., 3. 197.

18 Ronsard, *La Franciade / The Franciad*, 3. 225–226.

that patience is always a virtue, but that patience here makes sense as soon as Francus sees past his current frustration in the half-submerged ship, and out along the myriad connecting lines that reach from the ship back to Troy and forward to Paris. This kind of looking is ecological. Suspended in temporal as well as physical inertia, Francus must figure out how to deal with something bigger than he is, something elsewhere and elsewhen.

Rusty Advice

A final form with which the proverb collides in the *Franciade* is that of the aged advisor, as if the proverb were just such a being, almost geological in its ability to last through time and to remain pertinent for the present moment. At the close of *La Franciade's* fourth book, Hyante-as-prophet advises Francus to keep old men by his side as follows:

> Pource, Francus, si le ciel te fait Roy,
> Sage entretiens des vieillars prés de toy,
> Que te diront leurs raisons sans feintise
> En longs cheveux, en longue barbe grise.
>
> (Thus, Francus, if heaven makes you king,
> Be wise and keep old men at your side
> With long hair and long grey beards. They
> Will tell you their thoughts without disguise).[19]

Hyante may as well be saying 'Listen for proverbs!' and 'Keep an eye out for hyperobjects!' This emphasis on speech rather than action, on pausing to think with voices from other times and places, is structural within the *Franciade*, as becomes clear from the very beginning: the text opens not *in medias res,* as we expect from epic, but *in medias dictum* (i.e. with Jupiter's speech).[20] Moreover, the difference between Ronsard and Virgil here points to a similar emphasis. In the *Aeneid*, when Neptune calms the storm that

19 Ronsard, *La Franciade / The Franciad*, 4. 1627–1630.

20 In his 1572 preface, Ronsard admits that his decision to open the epic with Jupiter's speech is controversial: 'Je ne doute pas qu'on ne m'accuse de peu d'artifice en ce que la harangue de Jupiter au commencement de mon premier livre est trop longue, & que je ne devois commencer par là' ('I have no doubt that I will be accused of lacking craftsmanship regarding Jupiter's speech at the start of book one: it might be thought too long or the wrong place to begin'). Ronsard, *La Franciade*, p. 10 and *The Franciad*, p. 6.

threatens to put an end to the epic before it has begun, *this* is the work's inaugural decision and the first action taken on Aeneas's behalf. Neptune makes a short speech, chastising the winds for their disobedience, but Virgil makes sure that action wins out over speech: '*swifter than his word* [et dicto citius tumida aequora placat] he calms the swollen seas'.[21] By contrast, in the opening sequence of *La Franciade*, which narrates the council of the gods instead of the *Aeneid*'s storm, a simile is used to describe the gods consenting to Francus's epic destiny. This simile closely echoes Neptune's calming of the storm in the *Aeneid*:

> les Dieux qui s'esleverent,
> Tous d'un accord sa parolle aprouverent,
> En murmurant comme flots de la mer
> De qui le front commence à se calmer,
> Quand Aquilon assoupsist son orage,
> Et l'onde bruit doucement au rivage.

> (the gods stood up and
> Unanimously approved of his speech,
> Murmuring like the waves of the sea
> Whose surface begins to calm
> As Aquilo appeases the storm
> And the waves mumble softly to the coast).[22]

The gods of Ronsard's epic, amiably agreeing to let Francus have his destiny, transform Neptune's act of calming the sea into a mere simile for verbal acquiescence. Within this simile they also become the sea that is calmed—they are the passive matter of which Virgil's Neptune is the active principle. This conversion of action into speech is a doubled one. When Virgil's Neptune calms the storm, the effect is compared in an epic simile to that of a revered old man who calms a mutinous mob in the city: 'ille regit dictis animos et pectora mulcet' ('with speech he sways their passion and soothes their breasts').[23] That simile's *dictum* gets reabsorbed as one of the main events in *La Franciade*'s narrative: Jupiter prevents any mutiny amongst the gods (in particular Juno's mutiny) only by speaking to them; he is the old man calming the mob with his words. Both Neptune (in Virgil) and Jupiter (in

21 Virgil, *Aeneid*, 1. 142–143. Italics mine.
22 Ronsard, *La Franciade / The Franciad*, 1. 157–162.
23 Virgil, *Aeneid*, 1. 148–156.

Ronsard) step in to clear the way for their respective epic heroes, but whereas in the *Aeneid* action overtakes speech, in *La Franciade* the opposite occurs and, as a result, listening becomes a heroic quality.

Conclusions from Prince Dikaios and Erasmus

In her study of early modern commonplaces, Ann Moss argues that in the Renaissance these forms of speech are a 'store of quotations' that function as a kind of collective memory for the Renaissance humanist and which can 'be activated to verbalize present experience [...] with reference to a *cultural history shared by writer and reader*'.[24] As we have seen, such a definition holds true in the *Franciade* to a large extent—but only as long as we accept that that which is shared is never equally available to all. Proverbs offer not *just* a connection to that which is shared, but a self-aware rhetorical form. After his daughter Clymene becomes a casualty of Francus's stay in Crete, Prince Dikaios consults the oracle whose elliptical response is presented as a proverb:

> '*Si le Roy veut se soulager d'ennuy*
> *Ne loge plus d'arondelles chez luy*'.
> Telle parole en doute responduë
> Fut aisement de ce prince entenduë.
> C'est qu'il devoit par prudente raison
> Les etrangers chasser de sa maison',
>
> ('If the king wishes to be free of worry
> He must not lodge swallows in his house'.
> Such obscure words were easily understood by the prince.
> Reason and prudence required that he must
> Expel from his home all foreigners).[25]

Shadowing Virgil's Dido, Prince Dikaois immediately dismisses Francus as 'masqué de feintise' ('masked by deceit').[26] The word *feintise* will remind the careful reader that Francus was once indeed 'une feinte' ('a decoy')—that is, he was saved from death when Jupiter stole him, leaving a decoy in his place

24 Moss, *Printed Commonplace-books and the Structuring of Renaissance Thought*, p. vi.
25 Ronsard, *La Francaide / The Franciad*, 4. 15–20.
26 Ibid., 4. 33.

so that Pyrrhus would throw it and not Francus to its 'death'.[27] An invocation of 'prudente raison' ('prudent reason', which alludes to early modern debates about *prudentia* and *phronesis*) reminds us that Dikaios is a prince and must make the right sort of choice for his people. It is hard to deny that getting rid of Francus would be best for Prince Dikaios, since Francus seems about to make Hyante into his second (or third) Dido.[28] But just as Dikaios seems poised to ridicule Francus, saying that Francus operates 'soubs *couleur d'un destin*' ('cloaking himself in destiny'), he is pulled back.[29] The reversal is violent—and provides a final lesson about the ecology of proverbs that underwrites this epic:

> Que dis-je? Où suis je? En quelle folle erreur
> Perdant raison me pousse la fureur?
> *'Il ne faut pas qu'un prince debonnaire*
> *Du premier coup s'enflame de colere:*
> *Il ne doit croire aux flateurs de leger,*
> *Le commun bruit est toujours mensonger.*
> *Il doit attendre et sagement connoistre*
> *La verité que le temps fait paroistre'*
>
> (What am I saying? Where am I? To what mad
> Error does fury drive me as I lose
> My judgement? 'A gracious prince must not
> Be inflamed by anger on the first blow,
> He must not so lightly believe flatterers—
> Popular rumors are always lies.
> He must wait and learn wisely
> That truth born only from patience').[30]

At first blush this moment seems to signal Prince Dikaios's absolute rejection of the proverb as 'le commun bruit' ('popular rumours'). Patience is *opposed* to the proverb now, in stark contrast to Leucothea's earlier patience proverbs. However, this emphatic turn away from 'le commun bruit' is itself expressed in perhaps the most famous proverb in the entire epic: 'La verité que le temps fait

27 Ibid., 1. 108. On the significance of the word 'feinte' (decoy) see Usher's note in *The Franciad*, p. 31, n. 33.

28 See Francus seducing Hyante with some shameless lies: Ronsard, *La Franciade / The Franciad*, 4. 320–414.

29 Ronsard, *La Francaide / The Franciad*, 4. 35. For other articulations of this skepticism about epic destiny see Clymene's nurse (3. 1115–1136) and Phoveros's challenge to Francus (2. 1129–1143).

30 Ronsard, *La Franciade / The Franciad*, 4. 59–66.

paroistre' ('That truth born only from patience') is a version of the adage *veritas filia temporis* (truth is the daughter of time) that appears in Erasmus's famous collection, thus marking it immediately as a kind of an elevated proverb. This is in fact the proverb that justifies them all. Prince Dikaios is here staged realizing that the correct response to proverbs *is* a form of patience. 'Le commun bruit' of the oracular utterance can become the truth at an unpredictable pace (as Oedipus and Macbeth also discover). Prince Dikaios does not pull himself up short for listening to the oracle at all; rather he reminds himself that the time between listening to such messages and understanding them is crucial, and can be extended. Because proverbs are ancient speech acts that only gesture towards their content, and because that content is always at a remove and/or in retreat, listening to proverbs is more like putting your ear up against them and letting their 'empty' noise gradually resolve into a message that you can interpret and use. It is no coincidence that Dikaios has first to remember himself ('Que dis-je? Où suis je?': 'What am I saying? Where am I?') before he can realize his mistake. Standing at the interface between humans and the hyperobjects they live in and amongst, proverbs do not point humans towards the right answers but towards themselves in the very act of trying to figure things out. As Morton has put it, in the age of hyperobjects 'my intimate impressions' are no longer 'merely mine' or 'subjective only': 'they are the footprints of hyperobjects'.[31] Proverbs order and narrate human experience by giving us categories in which to place it. By linking singular experiences to radically undefined collectivities of experience, they offer humans a way out of the hopeless isolation of a given moment—if we make the effort to listen for a while. Prince Dikaios's final Erasmian utterance is his last contribution to the epic before he exits it for good, and *La Franciade* never reveals what truth the daughter of time delivers to him.

Works Cited

Braybrook, Jean, 'The Aesthetics of Fragmentation in Ronsard's *Franciade*', *French Studies*, 43.1 (1989), 1–11.

Burke, Kenneth, 'Literature as Equipment for Living' in *The Philosophy of Literary Form: Studies in Symbolic Action*, ed. by Kenneth Burke (Louisiana: State University Press, 1941), pp. 293–304.

Dionne, Craig, *Posthuman Lear: Reading Shakespeare in the Anthropocene* (Punctum Books, 2016).

31 Morton, *Hyperobjects*, p. 5.

Erasmus, Desiderius, *The Adages of Erasmus*, ed. by William Barker (Toronto: UT Press, 2001).
Hofmeyr, Isabel, and Kelly Bystrom, eds., *Oceanic Studies*, special issue of *Comparative Literature* 69.1 (March 2017).
MacPhail, Eric, *Dancing Around the Well: The Circulation of Commonplaces in Renaissance Humanism* (Leiden: Brill, 2014).
Morton, Timothy, *Hyperobjects: Philosophy and Ecology After the End of the World* (Minneapolis: University of Minessota Press, 2013).
Moss, Ann, *Printed Commonplace-books and the Structuring of Renaissance Thought* (Oxford: Clarendon Press, 1996).
Ronsard, Pierre de, *La Franciade*, in *Œuvres completes* vol. 16, ed. by Paul Laumonier (Paris: Marcel Didier, 1983).
———, *The Franciad (1572)*, ed. and trans. by Phillip John Usher (New York: AMS Press, 2010).
Usher, Phillip John, '*Non haec litora suasit Apollo*: la Crète dans la *Franciade* de Ronsard', *La Revue des Amis de Ronsard* 22 (May 2009), 65–89.
———, 'La Franciade', *The Literary* Encyclopedia (25 January 2011) http://www.litencyc.com/php/sworks.php?rec=true&UID=31708. Accessed 27 September 2019.
———, 'La Crète épique: La *Franciade* et la tradition des *isolarii*', *Cahiers V. L. Saunier* 34 (2017), 163–174.
———, *L'Aède et le géographe: Poésie et espace du monde à l'époque pré-moderne* (Paris: Classiques Garnier, 2018).
Virgil, *Eclogues. Georgics. Aeneid, Books 1–6*, ed. by G.P. Goold, trans. by H. Rushton Fairclough, Loeb Classical Library (Cambridge, MA: Harvard University Press, 2001)
Yaeger, Patricia, 'Editor's Column: Sea Trash, Dark Pools, and the Tragedy of the Commons' *PMLA* 125.3 (2010), 523–545.
Wofford, Susanne Lindgren, *The Choice of Achilles: The Ideology of Figure in the Epic;* (Stanford: Stanford University Press, 1992).

About the Author

Kat Addis is a doctoral student in the English Department at New York University, where she works on early modern literature. She is also a poet and an artist. Her work is published in The King's Review, ZARF, and Tears in the Fence (forthcoming). Recent film work is available online at kataddis.com. She is currently collaborating on projects with musician Mariam Bergloff and writer Joseph Minden.

5. Is Ecology Absurd? Diogenes and the End of Civilization

Pauline Goul

Abstract

This chapter proposes to unravel the many ecological underpinnings of Diogenes of Sinope's Cynicism. Perhaps thinking cynically about climate change requires going back to Ancient Cynicism in general, and Diogenes of Sinope in particular; within the argument of this volume, this chapter explores the resurgence of Diogenes and the particular tone of the works of François Rabelais and Michel Montaigne. It makes a convincing case for reading both of these authors less as polar opposites and more as thinkers of the ecological shift in early modern France.

Keywords: Diogenes, Michel de Montaigne, François Rabelais, Cynicism, humanism, cosmopolitanism

'The art of our necessities is strange
That can make vile things precious'.
Shakespeare, *King Lear*, III. 2. 70–71.

In *La Part du colibri* ('The Hummingbird's Share'), Pierre Rabhi, French farmer turned public environmentalist, writer, philosopher, and public figure, deplores what he calls the *névrose écologique*, to which he attributes the fact that so many people simply do not react in any significant way to the ecological crisis.[1] An Amerindian legend of the hummingbird frames this autobiographical essay on ecological becoming. Thousands of years ago, Rabhi narrates, an immense fire started in a great American forest, ravaging

1 Rabhi, *La Part du colibri*, p. 43.

Goul, P. and P.J. Usher (eds.), *Early Modern Écologies. Beyond English Ecocriticism.*
Taylor & Francis Group, 2020
DOI 10.5117/9789462985971_CH05

the land and leaving most animals to gaze, still and powerless, at the disaster ahead. A single hummingbird, however, could be seen transporting water in an attempt to appease the fire. An armadillo asked the hummingbird if it really believed it could stop the fire. 'I know, but I am doing my share', the hummingbird responded.[2] In the context of Western lives—urban or rural—can humans, like the other animals in the legend, reconcile the sense of the futility of their own actions with the need for more ecological lifestyles?

Often, we don't. In the face of climate change, the overwhelming response seems to be either plain denial or hopelessness: in other words, climate-scepticism or cynicism. Curiously, both reactions re-appropriate tone and language from ancient philosophy. Yet the climate-sceptics know little of Sextus Empiricus, and the climate cynics have rarely heard of Diogenes.[3]

This chapter will trace the strange resemblance between Diogenes—the scandalous, exiled philosopher who lived like a tramp in Ancient Greece—and modern figures of environmentalism like Rabhi, represented and stereotyped in the media for renouncing the comfort of modern civilization, living frugally on the outskirts of town, reprimanding the rest of humanity, sometimes even picking through the trash. If Cynicism has something to do with an approaching end, what can Diogenes bring to a consideration of ecology and of the end of the world? What would the infusion of Diogenic Cynicism into our modern ecological thought look like?

In fact, we find a proleptic response in the Renaissance. For it is the allusion to Diogenes of Sinope that, I will argue, marks an ecological thought in the work of French Renaissance writers Michel de Montaigne and François Rabelais. Scholars have begun to read Michel de Montaigne and François Rabelais environmentally,[4] and Diogenes enjoys renewed interest in recent scholarship on the French Renaissance, the philosophy of the Cynics, and the figure of Diogenes himself.[5] Despite his popularity, however, Diogenes is still often dismissed or misunderstood in the scholarship on the French

2 Rabhi, p. 10. Rabhi seems to be citing from somewhere, since the quotation is indented, but the origin of the legend is not indicated.

3 For the purpose of this chapter, the Cynics of Ancient Greece will take a capital 'C', while modern cynicism will be written with a small 'c'.

4 Including Phillip John Usher, Jennifer Oliver, and Antonia Szabari, to name a few.

5 See Michèle Clément, *Le Cynisme à la Renaissance* and Hugh Roberts, *Dogs' Tales: Representations of Ancient Cynicism in French Renaissance Texts*. An international colloquium on 'Ancient Cynicism and its Influence' in Paris in 1991 testified of this renewed interest, resulting in the publication of *The Cynics: The Cynic Movement in Antiquity and Its Legacy*, ed. R. Bracht Branham and Marie-Odile Goulet-Cazé. German philosopher Peter Sloterdijk also has a part in the revival of Cynicism with his *Critique of Cynical Reason*. Even more recently, a monograph by Etienne Elmer came out in French: *Diogène le Cynique*.

Renaissance. In the works of the writers I will study here, Diogenes is a central character in the important prologue to Rabelais's *Tiers Livre,* and he pops up here and there in Montaigne's *Essais.* Yet scholars frequently underestimate or downplay his significance. In *Dogs' Tales: Representations of Ancient Cynicism in French Renaissance Texts*, for instance, Hugh Roberts argues that Cynicism in the French Renaissance should be understood exclusively as a matter of humour and tone, 'as a commitment to serio-comic performance, including at times of war'.[6] I would, however, argue that the cynicism announced by the figure of Diogenes has much to do with the sense of an ending—of the world, of civilization—and that the serio-comic tone should thus be read as a crucial precursor to the modern concept of the absurd in theatre, which itself has a lot to do with the end of times and with ecology. Without misrepresenting the importance of Diogenes in particular and of Cynicism in general, I will argue that, in Diogenes, these writers find an ideal figure through which to express their environmental and existential dread. By tracing the representations of Diogenes in Michel de Montaigne's *Essais* and François Rabelais's *Tiers Livre*, and the many ways in which Diogenes resonates even when he is not identified, we will begin to see a thread of ecological cynicism linking these texts from Diogenes to Shakespeare's fools, to Samuel Beckett, and to our own sense of inefficacy in the face of the end times.

Tracing Diogenes in the French Renaissance

In the years since the publication of Michèle Clément's *Le Cynisme à la Renaissance* and Hugh Roberts's *Dogs' Tales*, the importance of Cynicism in French Renaissance texts hardly needs to be demonstrated. However, it is perhaps noteworthy that interest in the topic is relatively recent in the scholarship—Roberts and Clément published their books in 2005 and 2006 respectively. What is more, even scholars who engage the subject are cautious not to overstate its importance. Concerned with accuracy, Roberts, for instance, insists that 'Diogenes is not a key to Rabelais's work,' and describes readings that would suggest otherwise as being 'highly eccentric'.[7] His

6 Roberts, *Dogs' Tales*, p. 163.

7 He names Alice Fiola Berry, 'Apollo versus Bacchus: The Dynamics of Inspiration (Rabelais's Prologues to Gargantua and to the Tiers Livre)', *PMLA* 90.1 (1975), 88–95, and Florence Weinberg, '"A mon tonneau je retourne", Rabelais's Prologue to the *Tiers Livre*', pp. 548–563. For more details, see Roberts, *Dogs' Tales*, p. 172.

view, understandably, is that one should not grant too much importance to a character who only appears in the books a few times. We see emerging here a clear divergence between the treatment of the Cynics in Renaissance scholarship, which downplays or denies their socio-political importance, and their treatment in the Enlightenment scholarship, which acknowledges that Rousseau's and Diderot's versions of the Cynic were 'a template for a new breed of public intellectual'.[8] With Antónia Szabari, who asks us 'to entertain once again the idea of a Cynical Rabelais', I suggest that we consider anew the influence of the Cynics and of Diogenes in Renaissance texts, by examining the contexts and situations in which they arise.[9]

As Clément observes, the Cynics already appeared in various compilations in the late Middle Ages and continue to appear in the Renaissance. However, as she explains, the authors of Renaissance compilations had to deal with the fact that the public was now more familiar with the newly published original (for instance, Diogenes Laertius in 1533), which obliged writers to be more accurate in their descriptions of anecdotes. As a result, Diogenes operates a transition from a Christian ideal of asceticism to, in Clément's words, 'un Diogène nettement moins austère que précédemment, qui raisonne en joyeux pantagruéliste' ('a much less austere Diogenes than was previously thought, who reasons like a joyful Pantagruelist').[10]

Most scholars consider this ambivalence between Diogenes' asceticism and hedonistic tendencies to be irreconcilable; moreover, this could be the primary rationale for the reluctance to consider him a true, serious philosopher. It is, for instance, a major reason why Roberts refuses to consider Diogenes as doing anything other than setting a serio-comic tone in Rabelais's works. In his interpretation, Roberts chooses to only acknowledge the importance of Diogenes's joyfulness, completely leaving aside his asceticism: 'It is this aspect of Diogenes that Rabelais chooses to emphasize, and not his supposed Stoic indifference, nor his asceticism'.[11] However, in determining whether ascetic or hedonistic tendencies are more central, as R. Branham puts it, 'this project starts from the gratuitous assumption of the unity and coherence of Diogenes' thought'.[12] I would further argue that it is precisely to the extent that Diogenes can sustain such an ambivalence that Rabelais

8 Hershinow, 'Diogenes the Cynic and Shakespeare's Bitter Fool: The Politics and Aesthetics of Free Speech', p. 808.

9 Szabari, 'Rabelais Parrhesiastes: The Rhetoric of Insult and Rabelais's Cynical Mask', p. 84.

10 Clément, *Le Cynisme à la Renaissance*, p. 29. All translations from French scholarly works that have not been published in English are mine.

11 Roberts, *Dogs' Tales*, p. 176.

12 Branham, 'Defacing the Currency: Diogenes' Rhetoric and the *Invention* of Cynicism', p. 92.

can sustain his own: namely, the ambivalence between the performative excess of the first two books and the surprising call for moderation in the prologue for the *Quart Livre*, but always within a radical critique.

Who's afraid of Diogenes?

Fundamentally, Diogenes is the philosopher of appetites, a crucial concept in both Rabelais's narratives about over-consuming giants and Montaigne's *Essais*.[13] In his 2017 monograph on Diogenes, *Diogène le cynique*, Etienne Elmer frequently repeats the term *appétits*, insofar as, in Diogenes, it has a stronger meaning: he defines it 'au sens large d'exigences corporelles' ('in the broad sense of corporeal demands').[14] In Elmer's conception of Diogenes, the story begins with appetites, namely those of a mouse. In Diogenes Laertius, it is after having seen a mouse that Diogenes decides to live in a barrel: 'Through watching a mouse running about, says Theophrastus in the Megarian dialogue, not looking for a place to lie down in, not afraid of the dark, *not seeking any of the things which are considered to be dainties*, he discovered the means of adapting himself to circumstances'.[15] Indeed, the mouse has few appetites, and satisfies them quickly, for it is not weighed down by possessions or excessive needs. In the words of Elmer,

> Il signifie bien plutôt, et d'abord, qu'entre les appétits de la souris et leur satisfaction, l'intervalle est presque nul: elle court où elle se trouve, où qu'elle se trouve. Ce comportement de l'animal illustre ainsi sur un mode analogique ce que pourrait être une vie humaine simple, caractérisée non seulement par la réduction des désirs, mais aussi et peut-être surtout [...] par la réduction du nombre des médiations requises pour les satisfaire.[16]
>
> (He means, rather, and firstly, that between the appetites of the mouse and their satisfaction, the interval is almost non-existent: it runs to where it is, wherever it is. The behavior of the animal thus illustrates, on an analogical mode, what a simple human life could be, one that would be characterized not only by the reduction of desires, but also, and perhaps

13 See, for instance, in *Le Dictionnaire des Essais de Montaigne*, the article 'Désir (Appétit)', pp. 152–156.
14 Elmer, *Diogène le cynique*, p. 104.
15 Diogenes Laertius, *Lives and Opinions of Eminent Philosophers in Ten Books*, p. 25, my emphasis.
16 Elmer, *Diogène le cynique*, pp. 52–53.

> mainly [...] by a reduction of the number of mediations that would be necessary to satisfy those desires).

Cynicism, as it is expressed by Diogenes, is a philosophy of simplicity, of reducing the mediations that are necessary to satisfy human needs. In another of Diogenes Laertius's fragments, Diogenes declares 'that bad men obey their lusts as servants obey their masters'.[17] Appetites are to be controlled, not ignored. For Diogenes, in Ancient Greece, there are already too many of them and they are given free reign. It is primarily insofar as appetites are taken to signify any sort of demand from the body that Cynicism could be an ecological critique, already prepared to call out the expenditure of resources due to endless appetites.[18] It is thus by not simply contemplating appetites but also ascribing moral judgments to them, through the call for their reduction or simplification, that Cynicism becomes a mode of proto-ecological thought.

In fact, Diogenes' ecology spreads from the individual to the social, from the desires and appetites of the body to a striking premonition of capitalism: the growing distance of man from nature. Cynicism takes issue with superfluous objects; all would be well if man could be satisfied with fruits from his own garden. At the end of Laertius's sixth book, the fragments are devoted to a more general vision of the Cynics, written in the plural. One of them is an apt synthesis of the Diogenic ecological conception of life:

> They also hold that we should live frugally, eating food for nourishment only and wearing a single garment. Wealth and fame and high birth they despise. Some at all events are vegetarians and drink cold water only and are content with any kind of shelter or tubs, like Diogenes, who used to say that it was the privilege of the gods to need nothing and of god-like men to want but little.[19]

Elmer also points to fragment 44 to explain that 'La vie facile échappe aux hommes parce qu'ils recherchent des gâteaux de miel' ('Easy life is out of men's reach because they are searching for honeyed cakes').[20] In the original, Diogenes Laertius lists not only honeyed cakes but also 'unguents and the

17 Laertius, *Lives and Opinions*, VI. 66, 69.

18 I understand 'ecology' in the present chapter in its more modern meaning of, in the *Oxford English Dictionary*, 'The study of or concern for the effect of human activity on the environment; advocacy of restrictions on industrial and agricultural development as a political movement'.

19 Laertius, *Lives and Opinions*, VI. 105, 109.

20 Elmer, *Diogène le cynique*, p. 83.

like'.[21] This is usually the part of Diogenes' meagre doctrine that scholars and critics denounce as being anti-civilization or anti-promethean. As I will argue below, this is also the part that perhaps blinds Montaigne to how otherwise similar his and Diogenes' ideas are. The Cynics' frugality is often seen as a radical and gratuitous renunciation of the comforts of civilization. In this accusation, they evidently resemble modern environmentalists, from Henry David Thoreau, who lived on the edge of civilization in Walden, to Pierre Rabhi, who has a farm in the Cévennes and in 2015 inaugurated a sort of ecological vacation centre called *Le Domaine de l'Ermitage* (literally, the Hermit's Retreat). Rabhi's denunciation of capitalism and globalization even adopts a similar language in *Vers la sobriété heureuse*, a title that already sounds Cynic precisely for its juxtaposition of simplicity and joyfulness: 'La modernité ne serait-elle pas en train de gagner, insidieusement mais sûrement, la bataille de l'aliénation définitive de la personne, en la rendant dépendante des outils prétendant la libérer?' ('Is not modernity today insidiously but certainly winning the battle of the definitive alienation of the person, by rendering her dependent on the very tools supposed to liberate her?').[22] As an environmentalist, and despite Latour's *Nous n'avons jamais été modernes*, it is not surprising that Rabhi accuses modernity of a crime that Diogenes already assigned, centuries before the industrial revolution, to the distancing of man and nature by the former's appetites, and it could certainly lead us to further question the notion of 'Anthropocene' and its generally assumed modernity. Rabhi, moreover, seem either not to acknowledge or not to know that his vision of sobriety is steeped in centuries of a tradition of frugal philosophy, nourished by Cynicism. When he refers to cynicism, he speaks of the modern concept, accusing the system itself of being cynical.[23]

One can see how frugality could be the part of Cynicism to which Montaigne would grant the highest value. In the 'Apologie de Raimond Sebond', Montaigne rewrites or cites an anecdote centered around Diogenes washing vegetables. In Diogenes Laertius, this anecdote appears twice, the second time in paragraph 58 of Book VI, where Plato is the one telling Diogenes that

21 Laertius, *Lives and Opinions*, VI. 44, 47.

22 Rabhi, *Vers la sobriété heureuse*, p. 38.

23 Rabhi, *La Part du colibri*: 'Tout cela n'exclut pas non plus la production industrielle de biens, ajustés à un usage rationnel, solides, durables, modérés et non cette pléthore générant des rebus monstrueux dont la production massive est significative de l'inintelligence du système ou peut-être de son cynisme' (All this does not exclude the industrial production of goods, provided they are adjusted to a rational use, solid, durable and moderate, instead of the plethora that generates monstrous charades, the massive production of which reveals the lack of intelligence of the system, or perhaps its cynicism), p. 29.

he would not be washing vegetables if he had successfully flattered the king.[24] But it appears first in Book II on Aristippus (paragraph 68), where Diogenes is the one provoking his adversary, claiming that Aristippus would not need to flatter the tyrants if he had learned to live off of these vegetables.[25] In both cases, the anecdote ends on the facetious inversion of the first sentence, serving simultaneously as a criticism and an apology for either frugality or flattery. Aristippus retorts, 'And if you knew how to associate with men, you would not be washing vegetables'—accusing Diogenes of being a misanthropist and anti-civilization—whereas Diogenes retorts to Plato, 'Had you paid court to Dionysus, you wouldn't now be washing lettuces'—denouncing Plato's (admittedly false) praise of the powerful. It is the encounter between Aristippus and Diogenes that Montaigne relates, although it is juxtaposed with another anecdote about 'Dionysius le tyran' ('Dionysius the tyrant') and Plato's interaction with him, demonstrating Montaigne's conflation of several anecdotes. Significantly, the anecdote is also an addition from the *exemplaire de Bordeaux*. Yet in this instance, Montaigne does not cite it to call for frugality. Instead, he uses it to show that there are always two opposing ways of looking at things: 'Diogenes lavoit ses choulx, et le voyant passer: Si tu sçavois vivre de choulx, tu ne ferois pas la cour à un tyran. A quoy Aristippus: Si tu sçavois vivre entre les hommes, tu ne laverois pas des choulx. Voylà comment la raison fournit d'apparence à divers effects' ('Diogenes was washing his cabbages, and said, seeing him pass: If you knew how to live on cabbage, you would not pay court to a tyrant. To which Aristippus: If you knew how to live among men, you would not be washing cabbages. See how reason provides plausibility to different actions').[26]

Montaigne, however, is undeniably an advocate for frugality and moderation in the rest of his *Essais*. Indeed, like Diogenes, he criticizes the vulgar appetites of man, which conflate desires and appetites, the sexual and the comestible, in a similar way to Elmer's definition of appetites. In fact (and this shows that Montaigne's view of Diogenes and Cynicism is certainly problematic and ambivalent), he primarily judges them harshly for their open and free conception of sexuality.[27] On one occasion, he cites the 'embrasse-

24 Laertius, *Lives and Opinions*, VI. 58, 59–60.

25 Laertius, *Lives and Opinions*, II. 68, 197. In fact, a similar anecdote is also found, this time featuring Aristippus and another philosopher (Metrocles) in II. 102.

26 Montaigne, *Essais*, p. 581. Translation here and throughout the chapter are from *The Complete Works,* trans. by Donald Frame, p. 533.

27 Michèle Clément notes that Montaigne is mostly influenced by Diogenes Laertius's text, not Diogenes himself (p. 36). She adds that most of the topics Montaigne is interested in are Cynic

ments cyniques' (the 'licentiousness of the Cynic embraces'), 'mouvements lascifs' ('lascivious movements'), 'l'impudence de la profession de leur école' ('the shamelessness that their school professed'), and goes on to list their shameless sexual acts by describing Diogenes masturbating in public. In this way, Montaigne's opposition to Cynicism results from the philosophy not requiring enough limitations on those appetites: 'et n'ordonnoyent aux voluptez autre bride que la moderation et la conservation de la liberté d'autruy' ('and ordered no other bridle on sensual pleasures than moderation and the preservation of the liberty of others').[28] The point here is not to argue that Montaigne misunderstood and thus misrepresented the Cynics. Rather, I mean to insist on the fact that his use of Cynicism does not align with his own philosophy and praise of moderation. Furthermore, Montaigne seems to take issue with propriety and 'discretion', for in this anecdote he conflates the sexual debauchery of the Cynics and the fact that Diogenes 'mange[ait] qu'en pleine rue' ('would only eat out in the street'). From his tower, the dutiful Montaigne cannot forgive the Cynics for their criticism of the idea of a civilizing progress and their renunciation of the basic rules of civilization. Insofar as their philosophies are so similar, Montaigne's Diogenes is the proof of a missed encounter. His misgivings about the Cynics, that is, align with today's societal or the public opinion view of environmentalists; the prejudices and stereotypes held against their way of life prevent public opinion from asking whether their philosophy is worthy.

Rabelais's Prologue to the *Tiers Livre:* Diogenes' Political Ecology

If the ancient Cynics were to somehow reappear in 2019, they would probably be categorized—and perhaps dismissed—as radical ecologists. They are beggars, incontestably, but for economic and ecological reasons: they refuse the greed that is inherent in any sort of exchange and only accept donations or what they can procure by themselves. In this way, and for other reasons as well, the gleaners in Agnès Varda's *Les Glaneurs et la glaneuse* are Diogenic, although some more so than others, as I will explain below. Through their gleaning, they avoid the superfluous, which they seem to hold

topics: 'primat de la nature, coïncidence du bien et de la nature, méfiance envers la science, la défense des animaux' (the precedence of nature, the coincidence of goodness and nature, distrust of science, defense of animals) (p. 68) and asks the important question: 'Pourquoi Montaigne a-t-il tant de mal avec le mot "cynique"?' (Why did Montaigne have so much trouble with the word 'cynic'?), (*Le cynisme à la Renaissance*, p. 166).

28 Montaigne, *Essais*, p. 585; *The Complete Works*, p. 536.

in utmost contempt: it is the superfluous they criticize in the gibes at honey cakes and perfumes, but also in Diogenes' well-known rebuttal of Philip of Macedon, declaring himself to be 'a spy upon [his] insatiable greed'.[29] In one of Erasmus's *Apophtegmes*, Diogenes makes a similar observation of Alexander the Great: 'But Diogenes answered, which of us needs more things? I, who needs nothing else than my coat and my bag? Or you, who, not satisfied with your father's kingdom, expose yourself to so many dangers, in order to expand your empire further, in such a way that it seems that the whole world itself would not satisfy your enterprise and your desires?'[30] Greed may start with gluttony, but it evolves into expansionist politics and wars. Through this, Diogenes is already criticizing a form of globalization as a moral symptom, centuries before the first 'mondialisation' ('globalization') even occurs—the one Serge Gruzinski sees in the early modern Spanish empire.[31] Pierre Rabhi, a declared *altermondialiste* (alter-globalist), denounces the 'choix de l'antagonisme comme principe de vie' ('the choice of antagonism as a principle of life').[32] This is, for Elmer, the true sense of Diogenes' cosmopolitanism, often erased and omitted behind the veil of the accusations of being anti-civilization. I would argue that this is his political ecology, the best demonstration of which can be found in Rabelais's prologue to the *Tiers Livre.*

Rabelais's Diogenes has little to do with frugality and instead a lot to do with the often misrepresented and minimized politics of Cynicism.[33] His source for the anecdote is Lucian's *The Way to Write History.* The scene takes place in Corinth, and Rabelais stages the siege of the city by the already-mentioned Philip of Macedon. The inhabitants are busy preparing for the defence. Diogenes first observes them for a few days before deciding to join in the preparations by steadily rolling 'le tonneau fictil, qui pour maison luy estoit' ('the earthenware barrel that served him as a house') up and

29 Laertius, *Lives and Opinions*, VI. 43, 45.

30 Lobbes, *Des Apophtegmes à la Polyanthée, Erasme et le genre des dits mémorables*, p. 900. Note how close this is to some of Montaigne's main recriminations against imperialism and colonization in 'Of Coaches'.

31 Gruzinski, *Les Quatres parties du monde.*

32 Rabhi, *La Part du colibri*, p. 16.

33 Michèle Clément, for instance, apparently describes such an apolitical and asocial view of Cynicism as originating from a negative (perhaps prejudiced) interpretation of the Cynics' cosmopolitanism. See *Le cynisme à la Renaissance*, p. 64 : 'Ceux qui refusent une conception sociale et politique au cynisme, au nom de son intransigeance et de son individualisme, ont souvent considéré le cosmopolitisme cynique comme négatif' (Those who refuse the political and social conception of cynicism, in the name of its intransigence and individualism, have often considered the cosmopolitan dimension of cynicism as a negative thing.).

down a hill.[34] This summary only takes a few lines here, but the episode as recounted in the prologue occupies a few pages. Rabelais describes both the movements of the inhabitants and those that Diogenes inflicts on his barrel ('son tonneau ainsi tormenter') with a cornucopia of active, lively verbs, all very close in meaning in a characteristically Rabelaisian list. The prologue has been commented upon many times: scholars either read it as Diogenes mocking and satirizing the agitation around him, turning him into the misanthropic, anti-civilization version of the philosopher, or—at the other of the spectrum—they argue that it is his own way of participating. The latter is Weinberg's reading: 'the philosopher, under stress of coming combat, resolutely remained *au dessus de la mêlée*, although he manifestly did not scorn, and even symbolically joined, the activity surrounding him'.[35] Like his contradictory hedonistic and ascetic tendencies, scholars cannot reconcile the satirical dimension with a serious one. It is on this point that Roberts's main argument requires expanding. By giving a voice to the serio-comic, Diogenes illustrates that both the serious and the comic are crucial. Indeed, Branham argues that Cynicism is a philosophy of life according to nature, but only insofar as nature is unpredictable. The philosopher has to adapt himself to the situation:

> If Plato's paradigm is that of philosophy as *theōria* and the philosopher as a spectator of time and eternity, uniquely able to rise above time and chance, Diogenes' is just the opposite—the philosopher of contingency, of life in the barrel, of adapting to the *données* of existence, of 'minimal living', as Dudley puts it. On this view philosophy is not an escape from but a dialogue with the contingencies that shape the material conditions of existence.[36]

Diogenes' anecdotes are all situational. What Diogenes therefore offers by rolling his barrel up and down is an *alternative* to either fully participating in the preparations of war, or fully standing aside:

> Ce voyant quelqu'un de ses amis luy demanda, quelle cause le mouvoit, à son corps, son esprit, son tonneau ainsi tormenter? Auquel respondit le philosophe, qu'à aultre office n'estant pour la republicque employé, il

34 Rabelais, *Œuvres complètes*, p. 347, trans. by Donald Frame in *The Complete Works of François Rabelais*, p. 254.

35 Weinberg, '"A mon tonneau je retourne", Rabelais's Prologue to the *Tiers Livre*', p. 549.

36 Branham, 'Defacing the Currency: Diogenes' Rhetoric and the *Invention* of Cynicism' 89.

> en ceste façon son tonneau tempestoit, pour entre ce peuple tant fervent et occupé, n'estre veu seul cessateur et ocieux.
>
> (Seeing this, one of his friends asked him what cause impelled him thus to torment his body, his spirit, and his barrel. To which the philosopher replied that being employed on no other business by the commonwealth, he harried his barrel this way amid this people so fervent and occupied, not alone to seem a slacker and an idler).[37]

This description figures the Cynic in a composite way, illustrating key elements of the philosophy that will be important in determining its ecological dimension. First, Rabelais portrays the Cynic as a man both moved and on the move, far from the image of a lazy beggar who would just sit around near his barrel all day.[38] 'Seul cessateur et ocieux' then sounds like the usual accusations that would be formulated against Cynics, namely, that of being antisocial and apolitical. It also shows that Diogenes is consciously offering an alternative to what could be called the mainstream occupations of everybody else ('à aultre office', 'en ceste façon'). Moreover, despite stereotypes of the Cynics remaining outside of civilization, Rabelais emphasizes the fact that they always were in the middle of cities ('entre ce peuple'); if his barrel has no fixed location, Diogenes eats his cake, lives and interacts in the middle of the Agora.[39] All of these elements prove that he has a place—a share—in the public square and in the politics of the city, albeit an alternative one. Or, perhaps, *precisely* an alternative, and very active, place.

Diogenes moves considerably in the episode, at least as much as the sum of all the other inhabitants, creating an optical illusion, the effect of piling up active verbs for a long paragraph describing the preparations of war, which

37 Rabelais, *Œuvres completes*, p. 348; *The Complete Works,* p. 256.

38 According to Elmer, most of these accusations would actually be directed at disciples of the Cynics who would deform their teachings only for the sake of idleness: 'Pourtant, à côté de ceux qui le discréditent ou s'en prennent à ses disciples ignorants, crasseux et paresseux, de nombreux témoignages anciens voient aussi en lui le détenteur d'une profonde sagesse et d'une authentique philosophie, capable de déplacer et de réélaborer les grandes questions conceptuelles concernant l'homme et son rapport au monde, et vivant en accord avec ses principes' (Yet, next to those who discredit him, or criticize his ignorant, dirty and lazy disciples, numerous ancient testimonies see him as the bearer of a profound wisdom, of an authentic philosophy, both capable of displacing and reelaborating the great conceptual questions concerning man and his relationship with the world, and living according to his own principles.) : *Diogène le cynique*, pp. 16–17.

39 As specified in § 23, his barrel was located 'au Métrôon', that is to say, in Cybel's temple, east of the Agora. See Laertius, *Lives and Opinions*, VI. 23.

is immediately followed by a description containing almost as many verbs of Diogenes rolling his barrel down. It is, of course, illogical that a single man could make as much noise and agitation as a whole city of people. The discrepancy is quite visible. The first list consists of nine sentences, most of them starting with either 'les uns' or 'les autres'. Moreover, the verbs are just as much part of the list as the complements, since they are all transitive. In the second list, Diogenes is the only grammatical subject, and his barrel the only complement. The first list has many actions and many objects, the second has only one action, although repeated *ad infinitum*, and one object. Because all the verbs are intransitive, and all are synonyms for moving the barrel, the second list is much more pointless than the first one, and much more Rabelaisian as a result.

The second list thus hints at one key aspect of Diogenes' philosophy: endurance. Fundamentally, as Branham synthesizes it, Cynicism is based on discipline and self-sufficiency (*askesis* and *autarkeia*), which partly explains the physical ordeals to which Diogenes submits himself on a regular basis: 'L'été, il se roulait sur du sable brûlant, tandis que l'hiver, il étreignait des statues couvertes de neige, tirant ainsi profit de tout pour s'exercer' ('And in summer he used to roll in it over hot sand, while in winter he used to embrace statues covered with snow, using every means of inuring himself to hardship').[40] Rabelais's prologue fits with this depiction of Diogenes exerting himself, practicing his philosophy instead of merely reciting it—his reproach to Plato. Such acts may appear gratuitous, just like the environmentalists' willingness to endure voluntary deprivations of comfort. Diogenes living in a barrel, or throwing away his dish upon seeing a boy who eats lentils from a cup formed by his own hand, is equally *at odds* with society as are Varda's gleaners—or, to a lesser extent, the woman who reduces all her household waste to a quart-sized jar a year ('Zero Waste Home'), or the man who wants to limit his carbon footprint as much as possible ('No Impact Man')—insofar as these are quite privileged, urban endeavours, which I have criticized elsewhere, and which David Correia calls 'bourgeois primitivism'.[41]

40 Laertius, *Lives and Opinions*, VI. 23, 27.

41 Laertius, *Lives and Opinions*, VI. 37, 39: 'One day, observing a child drinking out of his hands, he cast away the cup from his wallet with the words: "a child has beaten me in plainness of living". He also threw away his bowl when in like manner he saw a child who had broken his plate taking up his lentils with the hollow part of a morsel of bread'. 'Zero Waste Home' is a blog and a book by Bea Johnson, *Zero Waste Home: the Ultimate Guide to Simplifying Your Life*. 'No Impact Man' is a blog by Colin Beavan, a book, (*No Impact Man: the adventures of a guilty liberal who attempts to save the planet, and the discoveries he makes about himself and our way of life in*

From Homelessness to Cosmopolitanism

While it never was a philosophy of indulgence and laziness, the Cynics, like environmentalists today, presented their way of life as only difficult and radical in appearance, arguing for its inherent easiness and simplicity. As Elmer summarizes it, 'Mais une fois qu'on s'y est suffisamment avancé, la vie cynique se révèle facile' ('But once one is sufficiently engaged into it, the Cynic life proves to be easy').[42] In the French made-for-tv documentary, 'Econome Sweet Home' starring Pierre Rabhi and other ecologist-citizens, the filmmaker follows families who live an ecological *alternative*.[43] One of the families, due to the inventiveness of the father, lives in an almost completely self-sufficient way thanks to solar panels, a vegetable garden, and a homemade water-mill in the river in their backyard. In these cases, as well as that of the 'Zero Waste Home' and 'No Impact Man', there is no doubt that ecology is its own occupation. In order to devote oneself to living self-sufficiently and ecologically within a Western, highly developed capitalistic society, one has to refuse to take part in the established exploitative conditions of production and of commodity circulation. The father in 'Econome Sweet Home' has given up his job, and one can assume the mother's salary from her job as a nurse is necessary to procure the goods that cannot be produced or constructed from within the autarky of their own home. Similarly, 'Zero Waste Home' is the project of a stay-at-home mother, while 'No Impact Man' is that of a freelance journalist and writer. Indeed, as these accounts attest, these projects required significant time, at least at their beginning stages, to figure out the logistics, make food from scratch, shop for local produce, or find stores that sell in bulk. In order to attain the simplicity they claim, the ways such ecological endeavours are represented paradoxically involves an undeniable amount of effort.

It is on this point that environmental living so often fails to make it into the mainstream. The more rustic and isolationist environmentalists are, the more they tend to be dismissed by polite society as irrelevant, extreme, radical or hippies. Rabhi lives in the Cévennes and owns a small farm that he runs according to ecological principles. Insofar as he also looks for an alternative and denounces the superfluous, his own lifestyle does not

the process, and a documentary. For a critique of the latter, see Correia, 'Degrowth, American Style: No Impact Man and Bourgeois Primitivism', pp. 105–118).

42 Elmer, *Diogène le cynique*, p. 92.

43 'Econome Sweet Home', Jean-François Méplon, broadcasted on France 5, 2016.

align with modern, urban life. The Cynics, however, resemble those urban ecologists, who try to sustain themselves in the middle of civilization. The Cynics' alternative conception of life according to nature—that is, to say, begging in the middle of the city—is an urban ecology of homelessness that is also a humanist cosmopolitanism. It resonates in many ways with Agnès Varda's *Les Glaneurs et la glaneuse*, in which many of the characters live in precarious situations or in poverty.[44] They turn a precarious situation into an alternative to capitalistic society. There is a generosity in Varda's documentary that is not present in other accounts of such behaviors. In *Redemption*, a HBO documentary on people who pick through the trash to recycle cans in New York City, for instance, the account of volitional precariousness as an alternative to the regular labour force is coupled with the dreadful competition among 'canners' and their daily struggle of making enough to be able to afford food or a home. In Varda's documentary, on the other hand, the gleaners do not simply exit the capitalist free market to find themselves in an equally cut-throat 'free' market. Rather, Varda's gleaners seem to relate differently to possession and need, and in turn to others. One urban gleaner, an undocumented refugee, brings home scavenged food and broken appliances. He and his host then cook all the food, and fix the appliances, in order to give them away to neighbours and friends in need.

Reconsidered in light of these modern, urban environmentalists, Diogenes' homelessness could be even more ecological and humanist than one might think. As Elmer emphasizes, the figure of Diogenes concerns people and their relationship with the world: 'Bien plus qu'un simple provocateur et qu'un critique iconoclaste, Diogène est avant tout le philosophe de la reconquête d'un rapport de proximité au monde et à soi-même, dont il fait le point d'ancrage d'une liberté et d'un bonheur pouvant nous rendre invulnérables' ('Even more so than a mere provocateur or an iconoclastic critic, Diogenes is first and foremost the philosopher who reconquers a close relationship to the world and to oneself, which he designates as the bearing point of both freedom and happiness, which in turn could render us invulnerable').[45] *Ecology* derives from Greek *oikos*, 'home', yet Diogenes' sense of the local, of proximity with the environment is an ecology without a home—Cynics describe themselves as *aoikos*, homeless.[46] In this way,

44 Another of Varda's films, *Sant toit ni loi*, could be another obvious example, but it is *The Gleaners and I* that presents a series of Diogenic figures. For a study of both films, see Stoekl, 'Varda and the Limits of Gleaning'.

45 Elmer, *Diogène le cynique*, p. 23.

46 Ibid., p. 143.

Diogenes' philosophy can help formulate a definition of ecology that does not rely on the assumption of having a fixed abode or on ownership of assets or goods. Illustrating the Cynic formula 'everything belongs to the wise', Elmer writes:

> Ce dernier vivant frugalement et simplement, sans autres biens que son maigre accoutrement, elle ne désigne pas tant une relation de propriété ou de possession qu'un rapport au monde où le sage est partout chez lui parce que ce tout, en un sens, dépend ou vient de lui: 'tout est *du* sage', dit littéralement le grec, le génitif signifiant ici une origine.[47]
>
> (The latter, living frugally and plainly, without any other possession than his meager clothing, it [the formula] does not refer to a relationship of property or possession. Rather, it is a way of relating to the world where the wise man is everywhere at home because this whole, in a way, depends and comes from him: 'everything comes from the wise man', says the Greek, literally, the genitive referring here to origin).

The locus of Cynicism is at once inherently local and cosmopolitan, without being global. It is sort of *altermondialiste*. Its movement is simultaneously lively and inexhaustible, like Rabelais's barrel, and its purpose unclear, especially in Rabelais's version of Diogenes.

Vanity and the Absurd

It is precisely on the question of purpose that Rabelais's Diogenes has something to say about ecology. All of these movements, and the endurance they involve, are so at odds with the rest of humanity that they appear futile and vain, whether one considers Rabelais's prologue to the *Tiers Livre* or the Diogenic anecdotes from Diogenes Laertius or Erasmus. Why on earth does Diogenes think his single dish is superfluous and that he needs to get rid of it? Similarly, is all of this recycling really doing anything to avert environmental disaster? Diogenes and modern ecologists, for most people, take it slightly too far. They are radical. Somehow, whether in Ancient Greece, in the French Renaissance or now, Diogenes perpetually sounds out of place. Branham mentions the anecdote of Diogenes eating in the agora, with a striking English translation of the syllogism that the latter presents as defense:

47 Ibid., p. 128.

If to breakfast is not 'absurd' [*atopon*],
It is not 'out of place' [*atopon*] in the agora.
To breakfast is not 'absurd' [*atopon*].
It is not 'out of place' [*atopon*] in the agora.[48]

In Goulet-Cazé's translation, she uses 'déplacé' ('out of place') in the four occurrences of *atopon* in the syllogism. *Atopon* is a key word of Diogenic Cynicism, and it expresses the fact that Diogenes is always 'at odds' with the rest of humanity. *Atopon* recalls *aoikos*. The relationship to place, to the world, is somehow unhinged, out of joint in Cynicism. The substance of this relationship can be translated, in modern English, as 'absurd', with all the associations that follow from it. If Cynicism is an alternative, it struggles to assert its own efficiency. *Atopon* is, indeed, altogether 'improper, perverse, unsuitable, amiss', perhaps even, in my interpretation, queer, because it challenges the logic of production and productivity, and because, in the words of Carla Freccero in *Queer/Early/Modern,* queer is 'a word used to designate that which is odd, strange, aslant'—that is to say, *atopon*.[49] It is in this way that Cynicism begins to look like modern ecology, and vice versa: the inherent difficulty and undeniable marginality of the Cynic way of life look—at least in appearance—like that of those environmentalists, regardless of whether they call themselves such or not, that choose to live differently, and it is precisely that which arises so many misgivings against it. How, indeed, can environmentalists *sell* what would now be necessary to curb global consumption: an austere, frugal, utilitarian way of life, after decades, arguably centuries, of over-indulgence? And even more so, how can they sell it if the rest of the world is not convinced of the efficacy of such arduous endeavours? In fact, ancient Cynicism and modern ecology both stumble when it comes to this sense of inefficacy and futility.

This absurd dimension, furthermore, is also visible in the French Renaissance representations of Diogenes. In Rabelais's prologue, Diogenes' barrel is often considered to be a pretext for Rabelais's interrogation about the necessity of his own work: 'ay pensé ne faire exercice inutile et importun, si je remuois mon tonneau Diogenic, qui seul m'est resté du naufrage faict par le passé on de Mal'encontre' ('I thought I would perform no useless and importunate exercise if I agitated my Diogenic barrel, which alone has been left me from the shipwreck incurred in the past at the lighthouse by the

48 Branham, 'Defacing the Currency: Diogenes' Rhetoric and the *Invention* of Cynicism', p. 94. See Laertius, *Lives and Opinions*, VI. 69.
49 Freccero, *Queer/Early/Modern*, p. 5.

Strait of Malencounter').[50] Rabelais admits to not knowing exactly what his purpose is, but asserts the enduring movement above anything else: 'Ainsi demeurera le tonneau inexpuisible. Il a source vive, et vene perpetuelle' ('Thus will the barrel remain inexhaustible. It has living spring and a perpetual vein').[51] In a way, Rabelais does not care about the inefficacy and futility of his own writing, provided it continues. The prologue could mainly be about energy and enthusiasm ('not to put out my utmost effort myself, and not to accomplish by it that little, my all, that I had left') in the face of what Rabelais refuses to participate in—war in the kingdom of France. Yet a farther reaching interpretation could be ventured once this prologue is considered in light of *Gargantua*'s. Already in this text, Rabelais betrayed a critical view of the agitation men usually fall into, when he writes, imitating Erasmus (who was well versed in Cynicism himself): 'Mais ouvrans ceste boyte: eussiez au dedans trouvé une celeste et impreciable drogue [...], deprisement incroyable de tout ce pourquoy les humains tant veiglent, courent, travaillent, navigent et bataillent' ('but, on opening the box, you would have found inside a heavenly drug beyond price [...] incredible disesteem for everything on account of which humans lie so awake, run, labor, sail, and fight').[52] The most precious drug in the 'Silene' box, at the end of the gradation, is one that would allow anyone to see the futility of such agitation. Like the Cynics, Rabelais despises those appetites which drive man to act.

Montaigne also betrays a similar criticism, and yet he appears to be most Diogenic when not even bringing up Diogenes. However, when he is at his most Diogenic, he sounds both like a Cynic and like an environmentalist: 'Je ne desavouë pas l'usage que nous tirons du monde, ny ne doubte de la puissance et uberté de la nature, et de son application à nostre besoing. [...] Je me deffie des inventions de nostre esprit, de nostre science et art, en faveur duquel nous l'avons abandonnée et ses regles, et auquel nous ne sçavons tenir moderation ny limite' ('I do not deny the use we derive from the things of the world, or doubt the power and fertility of nature and its application to our need. [...] I do distrust the inventions of our mind, of our science and art, in favor of which we have abandoned nature and her rules, and in which we know not how to maintain either moderation or bounds').[53] In 'Of Coaches', where he declares the ruin of the New World to have been caused by human greed and appetites, his tone takes a decidedly absurd

50 Rabelais, *Œvres completes*, p. 349; *The Complete Works*, p. 257.

51 Rabelais, *Œvres completes*, p. 351; *The Complete Works*, p. 259.

52 Rabelais, *Œvres completes*, pp. 5–6; *The Complete Works*, p. 3.

53 Montaigne, *Essais*, p. 766; *The Complete Works*, p. 704.

turn. The description is quite apocalyptic indeed, with 'cette mesme image du monde, qui coule pendant que nous y sommes' ('this very image of the world which glides along while we live on it').[54] Yet it also proves ambivalent:

> Comme vainement nous concluons aujourd'hui l'inclination et la decrepitude du monde par les arguments que nous tirons de nostre propre foiblesse et decadence [...]; ainsi vainement concluoit cettuy-là [Lucretius] sa naissance et jeunesse, par la vigueur qu'il voyoit aux espris de son temps, abondans en nouvelletez et inventions de divers arts.
>
> (As vainly as we today infer the decline and decrepitude of the world from the arguments we draw from our own weakness and decay [...]--so vainly did this poet [Lucretius] infer the world's birth and youth from the vigor he saw in the minds of his time, abounding in novelties and inventions of various arts).[55]

Montaigne's comparison subtly relies on the fact that, despite being faced with the same arguments, his contemporaries and himself concluded the decline and approaching end of their civilization, while Lucretius, in Ancient Rome, thought them proof of its birth and youth. These arguments, in 'Of Coaches', I read as the various appetites that bring on, indifferently, on the one hand, novelties and inventions and, on the other, the apparent necessity to expand the lands into the New World and exploit and ruin this new land and its inhabitants. Montaigne suggests that the end of the world merely depends on one's tone, as a matter of interpretation. Despite such ambivalence, he resorts to vanity and futility:

> Qui mit jamais à tel pris le service de la mercadence et de la trafique? Tant de villes rasées, tant de nations exterminées, tant de millions de peuples passez au fil de l'espée, et la plus riche et belle partie du monde bouleversée pour la negotiation des perles et du poivre: mechaniques victoires.
>
> (Who ever set the utility of commerce and trading at such a price? So many cities razed, so many nations exterminated, so many millions of people put to the sword, and down, for the traffic in pearls and pepper! Base and mechanical victories!)[56]

54 Montaigne, *Essais*, p. 908; *The Complete Works*, p. 841.
55 Montaigne, *Essais*, p. 908; *The Complete Works*, p. 841.
56 Montaigne, *Essais*, p. 910; *The Complete Works*, p. 844.

In other words, if people had not developed an appetite for pearls and pepper, or honey cakes and perfumes, they would not have travelled across the sea to expand their empires beyond the local. Montaigne's cosmopolitanism admires the *sauvages* and deplores the vices of his own civilization. In tone at least, Montaigne is not as distant from the Cynics as he seems to think.

Faced with their own versions of a crisis, whether it is the expected end of the world or various wars, Rabelais and Montaigne display the two faces of the absurd: they both call out the vanity of the situation and the futility of human actions, yet they differ in tone. Montaigne sounds more serious, more discouraged, Rabelais more jovial and comical. Depending on the director, Samuel Beckett's *Fin de partie* can be more entertaining and playful or more anxiety-inducing and dark. It is always, however, a little bit of both. The theatre of the absurd usually stages marginal figures who also recall beggars, notably for Beckett's *En attendant Godot*. Recently, scholars have endeavoured to write about the ecological dimension of the theater of the absurd—a dimension altogether omitted by Martin Esslin's canonical *The Theatre of the Absurd* in 1962. His conception of 'the human situation in a world of shattered beliefs' is the result of metaphysical anguish, itself a result of existentialism, the Second World War, Hiroshima and Nagasaki, and totalitarian regimes.[57] Yet, as Carl Lavery and Clare Finburgh assert in their introduction to *Rethinking the Theatre of the Absurd*, the Theatre of the Absurd has an 'ability to express an emergent sense of ecological and environmental anxiety that today has become so palpable and potentially catastrophic'.[58] If indeed, as they argue, 'the Theatre of the Absurd articulates an important ecological shift in human perception', I would argue that, at various moments in history, we can spot other such shifts, shifts that are also marked by the tone of the Absurd. Diogenes' philosophy in Ancient Greece and the works of Rabelais and Montaigne in the French Renaissance are such moments. They all lead to the Absurdist stage, just like they announce its dialogues. Lavery and Finburgh provide an ecological reading of the Absurd as conceptualized by Albert Camus in *Le Mythe de Sisyphe*, another Ancient Greek character (although this time from mythology) who endlessly rolls a rock up and down a hill. However, while Sisyphus performs his task in Hell as punishment, Rabelais's Diogenes, does it as an alternative to normally accepted

57 Esslin, *The Theatre of the Absurd*, p. 5.

58 *Rethinking the Theatre of the Absurd: Ecology, the Environment and the Greening of the Modern Stage*, ed. by Carl Lavery and Clare Finburgh, p. 1.

behaviour, a satirical one, full of purpose despite its uselessness. With this Diogenes, let us entertain the idea of a more positive and joyful, although dark and fierce, ecology.[59]

Learning to Die with Diogenes

Pierre Rabhi always comes back to the question that he claims is frequently asked of him: whether he is pessimistic or optimistic about the future of humanity, faced as it is with the ecological crisis.[60] Roy Scranton, in *Learning to Die in the Anthropocene*, pastiches Montaigne's famous chapter entitled 'Que philosopher c'est apprendre à mourir', in order to convince his readers that, despite the undeniable pessimism of the news about climate change and the already irreversible damage to the planet, there is to be a new optimism for our civilization. His arguments seem to owe a lot to Cynicism: 'From the perspective of many policy experts, climate scientists, and national security officials, the concern is not whether global warming exists or how we might prevent it, but how we are going to *adapt* to life in the hot, volatile world we have created'.[61] Emphasizing adaptation, Scranton recalls Branham's view of Cynicism as a 'philosophy of contingency' where adaptation and improvisation are crucial, helped as they would be by the exercising of endurance and the practice of self-sufficiency. Perceiving the tone of the absurd, he also satirizes the US Climate Summit: 'as if the world's leaders had been cast in a business-class version of Samuel Beckett's *Endgame*'.[62] It is precisely because whatever human beings could do now to stop climate change would be vain that Scranton is able to posit, instead of a call to arms, a call to thought (so to speak) or, in his words, to humanism: 'Over and against capitalism, we will need a new way of thinking our collective existence. We need a new vision of who "we" are. We need a new humanism—a newly philosophical

59 The editors of *Rethinking the Theatre of the Absurd* evidently also evoke Timothy Morton and his *Dark Ecology* as a possible lens through which to read the Theatre of the Absurd today. In my dissertation, 'An Ecology of Waste: Transtlantic Excess in Renaissance France', I demonstrate, through a reading of the nativity scene in Rabelais's *Pantagruel*, that Morton's 'dark ecology' could benefit significantly from considering the serio-comic and satirical dimension it seems to imply, for it omits dark humour as an influence, whereas a lot of the reactions one can have when faced with climate change and melting ice in the poles are, indeed, absurd laughter.

60 In *La Part du colibri*, p. 43. In *Vers la sobriété heureuse*, p. 115.

61 Scranton, *Learning to Die in the Anthropocene*, p. 17.

62 Ibid., p. 63.

humanism, undergirded by renewed attention to the humanities'.[63] The way that he depicts this humanist ecological thinker is completely on par with Cynicism: first, he describes him as 'the one who is willing to stop and ask troublesome questions, the one who is willing to interrupt, the one who resonates on other channels and with slower, deeper rhythms'. Later on, he brings up Sloterdijk who 'sees the role of the philosopher in the human swarm as that of an aberrant anti-drone slow-dancing to its own rhythm, neither attuned to the collective beat nor operating mechanically, dogmatically, deontologically, but continually self-immunizing against the waves of social energy we live in and amongst by perpetually interrupting its own connection to collective life'.[64] It would be difficult to not identify Diogenes in this aberrant, slow-moving interrupter, recalling *atopon* and the movements of Rabelais's prologue.

Diogenes meanders around the lives of eminent philosophers, around Montaigne's *Essais*, and in Rabelais's prologue to the *Tiers Livre*, more like a walking interruption than a provocation. He calls into question most of the things the rest of humanity takes for granted as necessary comforts. Redefining ecology in the process, he does not ask troublesome questions for the sake of the environment—that would be the normal, accepted notion of ecology—but instead, profoundly challenges the way people perceive their position and their role in the middle of the environment, by denouncing the always-proliferating mediations that, in a vicious circle, generate and are generated by a multiplication of human appetites. His is a human ecology of endurance and persistence, but also of joy. It is an urban ecology of sobriety, but also an out-of-place, rustic one. A Diogenic ecology would acknowledge its own shortcomings and inconsistencies, for it is not necessarily a coherent project precisely insofar as it persists as a humanist and ethical one. When Rabelais's Diogenes 'tormente' ('torments') his barrel, it is not a tragic act, but an absurd one, with all the serio-comic ambivalence it implies. Regardless of its declared inefficacy, Diogenes keeps rolling his inexhaustible barrel, doing his pointless but moral share, like the hummingbird in the legend, in the 'insigne fable et Tragicque comedie' ('singular fable and tragic comedy') of human life. Even if the end is nigh, rather than dismissing him and his ecological descendants, perhaps we should instead admire their resilience and, even while laughing at its possible futility, share this burden nonetheless.

63 Ibid., p. 19.
64 Ibid., p. 87.

Works Cited

Beavan, Colin, *No Impact Man: the adventures of a guilty liberal who attempts to save the planet, and the discoveries he makes about himself and our way of life in the process* (New York: Farrar, Straus and Giroux, 2009).

Boudou, Bénédicte, *Le Dictionnaire des Essais de Montaigne* (Paris: Scheer, 2011).

Branham, Robert, and Marie-Odile Goulet-Cazé, eds., *The Cynics: The Cynic Movement in Antiquity and its Legacy* (Berkeley, Los Angeles, London: University of California Press, 1996).

Clément, Michèle, *Le cynisme à la Renaissance d'Erasme à Montaigne* (Geneva: Droz, 2005).

Correia, David, 'Degrowth, American Style: No Impact Man and Bourgeois Primitivism', in *Capitalism Nature Socialism,* 23.1 (2012), 105–118.

'Econome Sweet Home'. Directed by Jean-François Méplon, broadcasted on France 5, 2016.

Elmer, Etienne, *Diogène le cynique* (Paris: Les Belles Lettres, 2017).

Esslin, Martin, *The Theatre of the Absurd* (Garden City, N.Y.: Anchor Books, 1969).

Freccero, Carla, *Queer/Early/Modern* (Durham: Duke University Press: 2006).

Gruzinski, Serge, *Les Quatre parties du monde* (Paris: Martinière, 2004).

Hershinow, David, 'Diogenes the Cynic and Shakespeare's Bitter Fool: The Politics and Aesthetics of Free Speech', *Criticism* 56.4 (Fall 2014), 807–835.

Johnson, Bea, *Zero Waste Home: the Ultimate Guide to Simplifying Your Life* (London: Penguin, 2013).

Laertius, Diogenes, *Lives and Opinions of Eminent Philosophers in Ten Books, Volume 2: Books VI-X,* trans. by R.D. Hicks, Loeb Classical Library 184 (Cambridge, Mass.: Harvard University Press, 1925 (1979)).

Lavery, Carl, and Clare Finburgh, eds., *Rethinking the Theatre of the Absurd: Ecology, the Environment and the Greening of the Modern Stage* (London, New York: Bloomsbury, 2015).

Les Glaneurs et la glaneuse, dir. by Agnès Varda (Ciné Tamaris, 2000).

Lobbes, Louis, *Des Apophtegmes à la Polyanthée, Erasme et le genre des dits mémorables* (Paris: Honoré Champion, 2013), vol. II.

Montaigne, Michel de, *Essais*, ed. by V.L. Saulnier (Paris: Presses Universitaires de France, 2004).

———. *The Complete Works,* trans. by Donald Frame (New York, London: Stanford University Press, Random House, 2003).

Morton, Timothy, *Dark Ecology* (New York: Columbia University Press, 2016).

Oliver, Jennifer H., *Shipwreck in French Renaissance Writing: The Direful Spectacle* (Oxford: Oxford University Press, 2019).

Rabelais, François, *Œuvres complètes* (Paris: Gallimard, 1994).

———. *The Complete Works of François Rabelais*, trans. by Donald Frame (Berkeley and Los Angeles, California: University of California Press, 1999).

Rabhi, Pierre, *La Part du colibri, l'espèce humaine face à son devenir* (la Tour d'Aigues: Editions de l'Aube, 2006).

———. *Vers la sobriété heureuse* (Arles: Actes Sud: 2010).

'Redemption', directed by Jon Alpert and Matthew O'Neill, HBO, 14 Oct. 2013.

Roberts, Hugh, *Dogs' Tales: Representations of Ancient Cynicism in French Renaissance Texts* (Amsterdam: Rodopi, 2006).

Scranton, Ron, *Learning to Die in the Anthropocene* (San Francisco: City Lights Books, 2015).

Sloterdijk, Peter, *Critique of Cynical Reason* (Minneapolis: University of Minnesota Press, 1987).

Stoekl, Allan, 'Varda and the Limits of Gleaning', *World Picture Journal* 5 (Spring 2011) 'Sustainability', http://www.worldpicturejournal.com/WP_5/Stoekl.html. Accessed 27 September 2019.

Szabari, Antonia, 'Rabelais Parrhesiastes: The Rhetoric of Insult and Rabelais's Cynical Mask', *MLN* 120.1 (January 2005), 84–123.

Usher, Phillip John, *Exterranean: Extraction in the Humanist Anthropocene* (New York: Fordham University Press, 2019).

Weinberg, Florence, '"A mon tonneau je retourne", Rabelais's Prologue to the *Tiers Livre*', *Sixteenth-Century Journal* 23.3 (1992), 548–563.

About the Author

Pauline Goul Assistant Professor of French Literature at The George Washington University. She specializes in early modern French literature and environmental criticism, and she has published pieces in the *Forum for Modern Languages Studies* and in volumes like *Global Garbage* and *French Ecocriticism*. She is currently working on a monograph tentatively entitled *The Anxiety of Waste: New World, Environment and Literature in Renaissance France*.

Part 2

Nature's Cultures

6. Between Nature and Culture: The Integrated Ecology of Renaissance Climate Theories[1]

Sara Miglietti

Abstract

'Climate theories' are often explained away in scholarship as pseudo-sciences irrelevant to the modern world, or as morally problematic forms of geographic determinism. This chapter instead argues that such theories still offer a valuable lens not only for understanding how early modern people conceptualized the relationship between human culture and nonhuman nature, but also for resituating ourselves with respect to this very same issue. Are we humans above and outside nature, or are we an integral part of it, caught in its dynamics and affected by its internal changes—including those resulting from our own agency? Three sixteenth-century authors (Le Roy, Bodin, La Framboisière) are here brought into dialogue with contemporary thinkers (Descola, Latour) to reappraise the 'integrated ecology' of nature and culture proposed by early modern climate theorists.

Keywords: climate theory, nature/culture, determinism, Jean Bodin, Loys Le Roy, Nicolas Abraham de La Framboisière

Introduction

Clarence Glacken's monumental overview of environmental ideas from Antiquity to the Enlightenment, published in 1967 but still an essential reference in the field, includes several chapters on what are often called

1 All translations in this chapter are mine unless otherwise noted.

Goul, P. and P.J. Usher (eds.), *Early Modern Écologies. Beyond English Ecocriticism.*
Taylor & Francis Group, 2020
DOI 10.5117/9789462985971_CH06

'climate theories' (*théories des climats, Klimatheorien, teorie dei climi*).[2] Such doctrines are centered on the idea that place and climate shape the body, mind, and character of human beings, influencing moreover the organization and development of human societies. In his book, Glacken explores several moments in the long tradition of climate theories, including their origins in ancient Greece (with authors such as Hippocrates and Aristotle), their medieval reception, and their presence in the early modern period, often thought to have represented their 'golden age'.[3] Between the sixteenth and eighteenth centuries, Glacken shows, climate theories reached an unprecedented level of visibility (they were somewhat ubiquitous) and complexity (they were put to many uses). While acknowledging their historical importance, however, Glacken claims that early modern climate theories represented a dead end on the path of intellectual development, and that '[i]t would be useless to claim that [these theories] contributed anything to understanding the relation of human cultures and their natural environment'.[4] Glacken argues that these theories 'could by no conceivable means lead to science', and therefore considers them to be of only limited interest today. At best, he suggests, they have the negative merit of revealing 'the inability of two millennia of accumulated lore to be of any real help in explanation'.[5]

The present chapter takes a rather different approach. Instead of asking whether these theories were more or less 'scientific' or whether they could, in Glacken's words, lead in any way to science as we presently conceive it, I shall look at climate theories as meaningful indicators of the ways in which people in the early modern period understood their place within the natural world. Renaissance climate theories, I argue, show us a worldview in which our own established divisions between nature and culture did not necessarily operate, or operated in different ways than they do nowadays. This is why such theories, 'pseudo-scientific' as they may seem today, may provide us with an unexpected resource for rethinking the problems that haunt our own relationship to the so-called natural world.

If it is true that the fundamental challenge of our time is to generate an integrated 'ecology of relationships' that would allow us to overcome the modern divide between man and nature, then it seems to me that

2 Glacken, *Traces on the Rhodian Shore*. On climate theories, see also Zacharasiewicz, *Die Klimatheorie in der englischen Literatur und Literaturkritik* and Pinna, *La teoria dei climi*.

3 Lestringant, 'Europe et théorie des climats'.

4 Glacken, *Traces on the Rhodian Shore*, p. 460.

5 Ibid., p. 446.

Renaissance climate theories have much to offer contemporary debates.[6] Building primarily on French examples, I will suggest that while these theories did not deny the existence of a nature/culture divide, they framed it in terms of a complex, porous, and mutually enriching relationship that has none of the rigid dualism so prevalent in Western modernity. In this sense, early modern climate theories will prove a helpful travel companion for rethinking the question recently raised by the French anthropologist Philippe Descola: how can we 'recompose nature and society, humans and nonhumans, individuals and collectives, in a new assemblage in which they would no longer present themselves as distributed between substances, processes, and representations, but as the instituted expression of relationships between multiple entities whose ontological status and capacity for action vary according to the positions they occupy in relation to one another'?[7] While Descola certainly did not have Renaissance climate theories in mind when he wrote these words (which he intended as a roadmap for a possible future), I hope to show in this chapter that it would be difficult to capture the essence of climate theory more effectively than do these lines.

Granted, this is not climate theory as we are used to seeing it described in scholarship. Climate theory is most often presented as a rigid system of environmental causality, coextensive and in fact even synonymous with the various forms of geographic determinism that would crop up in the nineteenth and twentieth centuries and that were often put in the service of dubious ideologies and pursuits.[8] In the last few years, however, this view has come under close scrutiny, most recently by Jean-Patrice Courtois in his compelling reappraisal of eighteenth-century theories of climates (notably those of Montesquieu, Hume, and Voltaire).[9] Enlightenment climate theories, Courtois shows, are fundamentally about relationships and correlations, rather than about causality and effects; about probability,

6 The notion of an 'ecology of relationships' is taken from Descola, *The Ecology of Others*, p. 5; Catherine Larrère similarly calls for an 'integrated ecology' (*écologie intégrative*) that views man not an entity 'external to nature' and standing in a relation of 'domination or opposition' to it, but as 'a geographical agent' whose action 'does not interrupt natural processes, but rather inscribes itself within them' ('Montesquieu et l'espace', p. 154). A classic (but not uncontroversial) account of the modern construction of the nature-culture divide can be found in Bruno Latour's *We Have Never Been Modern*. A comparable, though largely different, narrative is in Michel Serres's *The Natural Contract*.

7 Descola, *The Ecology of Others*, p. 5.

8 See Staszak, 'Nature et culture: des origines du "déterminisme géographique"' and Hulme, 'Reducing the Future to Climate'.

9 Courtois, 'The Climate of the *Philosophes* during the Enlightenment' and Courtois, 'Le Physique et le moral dans la théorie du climat chez Montesquieu'.

rather than about determinism; about transactions between mankind and nature, rather than about nature's crushing power on mankind. As I hope to demonstrate in this chapter, Courtois's remarks about Enlightenment climate theories apply equally well to Renaissance climate theories, if not to all climate theories in general.

In what follows, I focus on three significant representatives of climate theory in sixteenth- and early-seventeenth-century France, namely the humanist Loys Le Roy (c. 1510–1577), who taught Greek at the Collège royal; the jurist and political writer Jean Bodin (c. 1529–1596), best known for his influential theory of sovereignty; and the physician Nicolas Abraham de la Framboisière (1560–1636), who taught medicine at Reims and also served as personal physician to King Henry IV. While different in many respects, the climate theories developed by these authors partake of a common anti-deterministic impulse, as they all envisage multiple ways in which humans can shield themselves from climatic influence, including diet, music, and a liberal education. These authors also challenge the idea of a rigid dualism between mankind and nature by describing mankind as embedded in nature and nature as embodied in mankind, in a dynamic relationship that leaves ample room for the agency of both. If all of this is true, it seems necessary to abandon the traditional view of climate theory as a static system of geographic determinism grounded in a dualism between culture and nature. We should instead embrace a new view (derived from close reading of the texts themselves) of climate theory as a dynamic system of mutual correlations between multiple entities that are simultaneously natural and cultural, and that define each other's place in an interconnected universe.

The Cosmic Web: Humans and Nature in the Renaissance

When we think about mankind–nature connections in the Renaissance, we might want to set aside for a moment the term 'environment' which tends to come naturally to our modern mind. One of the problems with this word—aside from the fact that it did not exist (not, at least, in its current sense) in the period in question[10]—is that it encourages us to conceive of the mankind–nature relationship in terms of a dualism between two distinct entities, one of which (i.e. nature) surrounds the somewhat passive other (i.e. mankind). In the Renaissance, however, it was much more common to see the human species as part of a natural continuum that descended

10 Miglietti and Morgan, 'Introduction: Ruling "Climates" in the Early Modern World', p. 2.

from the celestial sphere of stars and planets into the so-called sublunary world, through various layers of reality that were all thought to share an essential ontological unity and to act reciprocally upon one another. In this view, humans are not *surrounded* by nature; they are part of nature, just as nature is part of them, making up their flesh, bones, humours, and vital heat through combinations of the same four elements that constitute all things in the sublunary world. In this sense, humans and nature are not distinct and opposing entities: humans are, quite literally, nature embodied; they are embedded in nature in a way that makes it impossible to distinguish neatly between the two.[11]

This view of the universe as an organic and interconnected whole can be traced back to classical antiquity, when, in the words of one scholar, 'the demarcation between human and environment was only faintly drawn'.[12] A very similar view is reflected in many Renaissance texts, including Loys Le Roy's influential treatise *De la vicissitude ou variété des choses en l'univers* (1575), which was translated into Italian in 1585 and into English in 1594.[13] The enduring interest of this treatise lies less in its (rather limited) originality than in its ability to synthesize an entire worldview in twelve, neatly ordered books. In this work, Le Roy deals with many different topics—from the historical development of languages and arts to political institutions and military matters—in order to demonstrate the central idea (itself not especially original in the Renaissance) that the universe is 'temperé par changements alternatifs, et maintenu par contraires, demourant en son essence eternelle tousiours mesme et immuable' ('tempered by alternative changes, and maintained by contraries, its eternal essence remaining always one and unchangeable').[14]

One part of the treatise is particularly pertinent here, namely a section in Book 1 in which Le Roy describes the structure of the universe, drawing liberally from a longstanding cosmological tradition that had found its most

11 As Floyd-Wilson and Sullivan have noted, 'body and environment do not merely mirror each other' in early modern views of the mankind-nature relationship, 'they also interpenetrate' ('Introduction: Inhabiting the Body, Inhabiting the World'), p. 2.

12 Floyd-Wilson, *English Ethnicity and Race in Early Modern Drama*, p. 29. Floyd-Wilson refers in particular to the Hippocratic treatise *Airs, Waters, Places* (5th century BC), a foundational text of classical climate theory.

13 Le Roy, *De la vicissitude ou varieté des choses en l'univers*; and *Of the Interchangeable Course, or Variety of Things in the Whole World*, trans. Ashley. The English translation may have influenced Francis Bacon's theory of vicissitude, as suggested by Weisinger, 'Louis Le Roy on Science and Progress'.

14 Le Roy, *De la vicissitude ou varieté des choses en l'univers*, 1r; *Of the Interchangeable Course*, 1r. Here and elsewhere, translation modified.

concise and influential expression in Johannes de Sacrobosco's *Tractatus de sphaera* (*Treatise on the Sphere*), composed in the first half of the thirteenth century but still extremely widespread in the Renaissance.[15] Citing the views of 'most astrologers and physicians', Le Roy explains that:

> de la partie superieure de l'univers descen[d] certaine vertu accompagnee de lumiere et chaleur qu'aucuns d'eux appellent l'esprit de l'univers, les autres nature, se meslant parmy la masse de son grand corps penetrant, vivifiant, nourrissant, moderant toutes choses sublunaires variables. Laquelle estant de telle efficace commence au feu et à l'air, lesquels agitez par mouvemens coelestes, esmeuvent apres l'eau et la terre, consequemment les natures composees de ces quatre elemens tant hommes, bestes, poissons, oyseaux, que germes, plantes, arbres, pierres et metaux.[16]
>
> (From the superior part of the world there descends a certain virtue accompanied with light and heat, which some of them do call the spirit or soul of the world; others say it is nature, which mingles itself with the mass of this great body, penetrating, quickening, nourishing, and moderating all these variable things under the moon, which being of such efficacy, begins first with the fire and the air, which being moved by the celestial movings, do afterwards move the water, and the earth, and consequently the natures compounded of these lower elements, as well men, beasts, birds, and fishes, as plants, trees, herbs, and metals).

Two things are especially worthy of note in this passage. On the one hand, Le Roy brings out the notion of a chain of being—a descending hierarchy of living forms, all of which are connected in some way to each other, and which fill up the order of nature completely through their plentiful variety.[17] On the other hand, Le Roy draws attention to the ontological unity of the sublunary world by stressing that everything that exists on Earth or in its immediate surroundings participates of the same essential nature, resulting as it does from different combinations of the same four elements (fire, earth, air, and water). These two aspects are brought together in the following pages, where Le Roy delves deeper into the idea that everything in

15 Thorndike, *The Sphere of Sacrobosco*; Gingerich, 'Sacrobosco as a Textbook'; Valleriani, 'The Tracts on the "Sphere"'.

16 Le Roy, *De la vicissitude*, IV; *Of the Interchangeable Course*, IV.

17 For a classic study on this notion of a chain of being, see Lovejoy, *The Great Chain of Being*. A brief discussion of Lovejoy's ideas can be found in Glacken, *Traces on the Rhodian Shore*, pp. 5–6.

the cosmos is tied together by a web of influences and interconnections.[18] Again, the thought itself is not especially original—it was, as Le Roy himself acknowledges, a rather commonplace idea among Renaissance astrologers and physicians, whose respective disciplines were much closer to each other in the premodern period than they would become later. Both dealt with the study of the various factors (such as stars, planets, air, and food) that were believed to exert an influence on human bodies and minds:

> [Les astrologiens et physiciens affirment] de là proceder diverses temperatures des corps, inclinations d'entendemens, moeurs des personnes, proprietez des nations, vices et vertus, santé et maladies, force et foiblesse, brieveté et longueur de vie, mortalité: richesse et pauvreté, prosperitez et adversitez. De là prendre commencement les estats et sectes, leurs progrez, durees et ruines. Brief tout ce monde inferieur obeïr au superieur et par luy estre gouverné.[19]

> ([The Astrologers and Philosophers affirm] that there hence do proceed diverse temperatures of bodies, inclinations of minds, manners of men, properties of nations, vices and virtues, health and sickness, force and feebleness; shortness and length of life, mortality, riches and poverty, prosperity and adversity. That there hence all estates and sects do take their beginnings, their course, continuance, and their ends: In brief, that all this inferior world does obey the superior, and is governed by it).

What is interesting about Le Roy's stance is the decisiveness with which he affirms that the various external influences that condition the fate of individuals and communities do not, however, wield an absolute power over human beings:

> Non pas que tels effects adviennent necessairement et inviolablement par une loy fatale: ains qu'ils peuvent estre evitez par sagesse, ou destournez par prieres divines, ou augmentez et diminuez par prudence, ou moderez par nourriture, coustume, institution.[20]

> (Not that such effects do necessarily come to pass, and inviolably by a fatal law: but that they may be avoided by wisdom, or turned from us by

18 See Severini, '*La vicissitudine o mutabile varietà delle cose*', p. 140.
19 Le Roy, *De la vicissitude*, 1v; *Of the Interchangeable Course*, 1v-2r.
20 Le Roy, *De la vicissitude*, 1v; *Of the Interchangeable Course*, 2r.

divine prayers, or augmented or diminished, or moderated by nurture, custom, and instruction).

While Le Roy is far from being alone among his contemporaries in conceiving of celestial and elemental influences in non-deterministic terms (a favourite expression at the time was that such forces 'incline, but do not necessitate'), the care and precision that he displays in his choice of words is worthy of note. Le Roy lists a number of possible ways in which humans can cope with the various celestial and elemental influences acting upon them: these ways of coping range from evasive strategies aimed at avoiding environmental influences altogether (*evitez par sagesse, destournez par prieres divines*) to corrective practices (*nourriture, coustume, institution*) that allow man to 'moderate' or 'reduce' (*moderez, diminuez*) the effects of environmental influences—or even, in certain cases, to augment them artificially (*augmentez* [...] *par prudence*). Although Le Roy does not go into great detail when explaining how each of these different strategies is expected to work in practice, his rich vocabulary of coping clearly testifies to the non-deterministic spirit of early modern climate theories. As we shall see in the next sections, the 'anxiety of influence' elicited by these theories was never such that it led people into fatalism or despair; on the contrary, it encouraged people to assert their autonomy even more strongly in the face of external forces, and to fashion themselves as self-determining moral subjects through a range of individual and collective practices.

'Second nature': The Power of Corrective Discipline

This particular tension between influence and autonomy in early modern climate theories emerges with greatest clarity in the works of Jean Bodin, one of the most important climate theorists of all time. Bodin's *Methodus* (1566) and *République* (1576) are often described, with good reason, as true *summae* on this topic, for in them Bodin draws up the most systematic overview of climate theory ever attempted since the time of Albertus Magnus (thirteenth century), bringing together a wealth of ancient, medieval, and coeval sources (such as Leo Africanus's and Francisco Alvarez's descriptions of Africa), weighing discrepancies and contradictions between these sources, and striving to generate a coherent system out of them.[21] Bodin's survey is

21 See, for instance, Glacken, *Traces on the Rhodian Shore*, p. 434; Lestringant, 'Europe et théorie des climats', p. 206.

especially remarkable for its extensive coverage of different types of environmental influence: in studying the effects that climates and places have on human beings, Bodin considers aspects such as latitude and longitude ('at first we shall explain the nature of peoples who dwell to the north and to the south, then of those who live to the east and to the west') but also more specific features of the land and of its local climate ('next, we notice the characteristics of special places, that is, mountains, marshes, windy and placid regions'),[22] which enables him to form a more complete view of environmental influence than is the case with many of his contemporaries.[23]

Like Le Roy, Bodin is convinced that these external influences are powerful but not insurmountable: it is false, he writes in the *Methodus*, 'that the constitution of the air affects us inevitably [...]. Regions and celestial bodies do not have so much power as to entail necessity (which it is a sin even to imagine)'.[24] He does however point out that fighting environmental influence requires a considerable amount of self-discipline and will-power. Similar to Le Roy, who singles out prayer and education (*nourriture, coustume, institution*) as possible ways of coping with climate, Bodin thinks that the influence of environmental factors cannot be overcome 'except through divine aid or continued discipline' (*nisi ope divina, aut diuturna disciplina*).[25]

At first glance, Bodin's notion of corrective discipline may seem to point to a significant hiatus between nature and culture—more significant, it would seem, in Bodin's climate theory than in Le Roy's. Bodin himself tempts us into thinking that it was precisely this hiatus that attracted him towards climate theory in the first place. It is important to remember that much of Bodin's *Methodus* is concerned with the problem of how to reduce human history to order—how to find a rationality in the apparent chaos of human matters.[26] Several chapters in the treatise tackle this problem from a range of different perspectives: reading and note-taking strategies (Chapter 3); astrology (Chapter 5); numerology, vicissitudinal theory, and the comparative history of political institutions (Chapter 6); prophecy and sacred history (Chapter 7); chronology (Chapter 8); etymology and historical

22 Bodin, *Methodus ad facilem historiarum cognitionem*, 5.3, 220; Bodin, *Method for the Easy Comprehension of History*, trans. Reynolds, p. 85. Latin: 'primum igitur explicabimus naturam populorum qui ad Septentriones et Austrum positi sunt: deinde eorum qui ad ortum et occasum: post etiam propria loca, montana scilicet, palustria, ventosa, quieta'.

23 For the existence of different 'levels' (cosmological v. chorological) in climate theory, see Miglietti, 'New Worlds, Ancient Theories: Reshaping Climate Theory in the Early Colonial Atlantic'.

24 Bodin, *Methodus*, 5.4, 222; *Method*, p. 86.

25 Idem.

26 See Couzinet, *Histoire et méthode à la Renaissance*.

linguistics (Chapter 9); bibliography (Chapter 10). Some of these avenues (astrology, for instance) are assayed with a certain degree of skepticism; some are criticized and ultimately dismissed.[27] Climate theory, on the other hand, must have appeared to Bodin to be a particularly promising route, seeing as he not only devotes an entire chapter of the *Methodus* (Chapter 5) to this topic, but also returns to it ten years later in the *République* (Book 5, Chapter 1), where he modifies certain aspects but retains the gist of what he had proposed in the earlier work, while also further developing his theory of climates in the direction of practical governmental applications.[28]

The reason why climate theory proves so helpful is spelled out at the outset of Chapter 5 in the *Methodus*, where Bodin explains that this theory allows us to identify 'characteristics drawn, not from the institutions of men, but from nature, which are stable and are never changed unless by great force or continued discipline, and even if they have been altered, nevertheless eventually they return to their pristine character'.[29] This passage establishes a series of dichotomies between nature (*natura*) and culture (*instituta*): nature is that which remains stable, culture is that which changes over time; nature is that which is given, culture is that which results from human artifice (*quae ducuntur ab hominum institutis*); nature is that which comes first in order of time and is therefore more powerful, culture is that which comes second and is therefore less powerful and doomed to fight for its own existence. The passage also suggests that nature's dominant position over culture is the whole reason that climate theory can help us identify a hidden order in human matters: it is precisely because culture springs up in reaction (and therefore in relation) to nature that the study of natural circumstances can illuminate the study of cultural phenomena.

It is no wonder that this passage has been read as proof of Bodin's determinism: it does appear as though Bodin is pointing here to a radical divide between nature and culture, with nature taking precendence over culture while culture is reduced to a secondary and rather precarious role. Yet this is

27 See, for instance, his criticism of the Protestant interpretation of the prophecy of the four monarchies in the Book of Daniel, studied by Suggi, 'Cronologia e storia universale nella *Methodus* di Jean Bodin'.

28 For differences between the *Methodus* and the *République* with particular respect to climate theory, see Staszak and Couzinet, 'À quoi sert la "théorie des climats"?' and Spavin, 'Jean Bodin and the Idea of Anachorism'. Spavin's analysis also takes into account one of Bodin's later works, the *Universae naturae theatrum*, published in 1596.

29 Bodin, *Methodus*, 5.2, 220; *Method*, p. 85. Latin: 'illa quae non ab hominum institutis, sed a natura ducuntur, quaeque stabilia sunt, nec umquam nisi magna vi, aut diuturna disciplina mutantur; et mutata nihilominus ad pristinam redeunt naturam'.

not Bodin's final word on the matter. Everything that follows in Chapter 5 of the *Methodus*, as well as Bodin's later discussion of the topic in the *République*, call into question both the dualism and the hierarchy that are seemingly established here. First of all, Bodin is keen to stress that the effects of corrective discipline, while fragile and precarious, are real and proven by famous historical examples: the Arcadians, who used music to overcome the negative influences of their mountainous environment;[30] the Germans, who were able to rise from barbarity to civility through self-discipline and education;[31] and the Carthaginians and the Arabs, 'soft' southern peoples who nevertheless managed to establish powerful empires thanks to rigorous military training.[32]

Furthermore, the primacy of nature over culture posited at the outset of *Methodus* 5 is explicitly reversed in the *République*, where Bodin states unequivocally that 'nurture is stronger than nature' (*nourriture passe nature*), referring once again to the Germans as a case in point.[33] While in the *République* as in the *Methodus* Bodin is careful to stress that the corrective effects of discipline only last if the discipline itself is rigorously practiced ('it is true that if the laws and customs are not well maintained, the people will soon return to its natural character'),[34] Bodin does seem to uphold a more optimistic view in the *République* than he did in the *Methodus* regarding our ability to withstand the influence of climate. This increasing optimism is further signalled by the introduction, in the second revised edition of the *Methodus* (1572), of a passage on the Scythian Anacharsis, a philosopher born and bred in a northern climate (an unlikely home for intellectuals, according to the Mediterranean-centered outlook of classical climate theory).[35] This passage, absent from the first edition of the *Methodus*, presents Anacharsis as 'proof that [air] has indeed great influence for changing character, yet does not entail necessity',[36] further evidence indeed of the non-deterministic spirit of Bodin's climate theory.

30 Bodin, *Les Six livres de la République*, 5.1, 692.

31 Bodin, *Methodus*, 5.180, 332–334; *Method*, p. 145.

32 Bodin, *Methodus*, 5.181, 334; *Method*, p. 145.

33 Bodin, *Les Six livres de la République*, 5.1, 695. French: 'Mais qui voudra voir combien la nourriture, les loix, les coustumes ont de puissance à changer la nature, il ne faut que voir les peuples d'Alemagne, qui n'avoient du temps de Tacite ny loix, ny religion, ny science, ny forme de Republique, et maintenant ils ne cedent point aux autres peuples en tout cela'.

34 Ibid., 5.1, 695. French: 'vray est que si les loix et coustumes ne sont bien entretenuës, le peuple retournera bien tost à son naturel'.

35 On this topic, see Floyd-Wilson, *English Ethnicity and Race*, esp. Chapter 1.

36 Bodin, *Methodus*, 5.4, 222; Bodin, *Method*, p. 86. Latin: 'magnam quidem vim ad immutandos animos habere, necessitatem tamen non adferre argumento fuit Anacharsis Scytha'. Anacharsis was mentioned once in the first edition of the *Methodus*, in similar but much less explicit terms:

More radically still, the very dichotomy that opposes culture and nature apparently established at the outset of *Methodus* 5 is undone elsewhere in various ways. One particularly interesting case occurs later in the same chapter, where Bodin introduces a vegetal metaphor to reflect about the power of education:

> quemadmodum foecunda tellus nisi excolitur, magnam nocentium herbarum vim profert: et modice culta valde frugifera fit; sterilis vero neque salutares, neque noxias herbas, nec quicquam omnino nisi maximo labore parit: ita quoque de Australium ac Scytharum ingeniis iudico.
>
> (As the fecund earth produces a large supply of noxious weeds unless it is cultivated [*nisi excolitur*] and when worked [*culta*] in a proper manner becomes really fruitful; and sterile earth, on the other hand, produces neither healthful nor noxious weeds, nor anything at all except with the greatest effort; so also I judge to be the case with the talents of the southerners and of the Scythians).[37]

Here, Bodin deliberately plays on the polysemous word 'nature'—which can indicate both the physical world (what some call 'environment' today)[38] and the fundamental constitution of a thing (its 'nature' or essence)[39]—in order to problematize the dichotomy between *natura* and *instituta* with which he had opened the chapter. The opposition sketched in this passage is not so much between nature (intended as environment) and culture (intended as the world of humans), but rather between a nature that can be improved by culture on the one hand, and a nature that is infertile, and therefore intractable to culture, on the other. By establishing this opposition, Bodin draws attention to the fact (already highlighted by Cicero in his *De natura deorum*) that most of the time the physical nature that humans experience is not a pristine wilderness but a 'second nature' already modified by culture (*culta*) in more or less visible ways.[40]

'nullos unquam a Scythia philosophos praeter Anacharsim; innumerabiles a Graecia fluxisse' (5.84, 272–274).

37 Bodin, *Methodus*, 5.74, 268; Bodin, *Method*, p. 110.

38 See *OED*, 'Environment', 2d.

39 See Williams, *Keywords: A Vocabulary of Culture and Society*, p. 220, on the polysemy of the word 'nature'.

40 For a discussion of Cicero's concept of 'second nature' and its influence on early modern theories of the landscape, see Hunt, *Greater Perfections*.

Bodin's use of the vegetal metaphor further challenges any rigid nature/culture dichotomy by suggesting that humans themselves do not belong exclusively in the realm of culture—they are also firmly anchored in the realm of nature: they *are*, in fact, nature, so long as culture does not intervene to complicate this identity. The point is made most clearly later in the chapter, where Bodin brings together Cicero's concept of 'second nature' with the Aristotelian notion of *hexis* or *habitus* (a stable disposition acquired through long habit), writing that 'such is the influence of custom and discipline in natural and human affairs that gradually they develop into into mores and take on the force of nature'.[41] In other words, culture itself can become nature by means of constant repetition (*diuturna disciplina*). Like fertile lands, then, humans are for Bodin fundamentally in-between nature and culture: simultaneously exposed to environmental influences, and capable of mastering these influences to a certain extent through *cultura* ('culture', but also 'cultivation'). It seems then that the ultimate goal of Bodin's climate theory is not to construct a hierarchical dichotomy that separates nature and culture, but rather to draw attention to the interstitial spaces *between* nature and culture, so as to illuminate their mutually-constitutive relationship.

A Balancing Diet: The Medical Economy of Climatic Influence

Bodin's corrective discipline, as we have seen, can take many forms: from religion and music to laws, military training, and an education in the liberal arts. However, one important dimension that is missing from Bodin's discussion is that of food and diet as countermeasures against climatic influence. While Bodin does speak at length about the relationship between food and climate in *Methodus* 5 (and to a lesser extent in *République* 5.1), he does not go into great detail to explain how diet fits in his picture of corrective discipline. Unsurprisingly, this aspect takes centre stage in discussions of climate theory by professional physicians and dietitians. The idea of responding to the influence of climate through diet is not specific to the Renaissance: it dates back to ancient medical writers such as Hippocrates and Galen, as well as to other authors (Plutarch, for instance) whose dietetic advice was steeped more in moral philosophy than in medicine proper.[42]

41 Bodin, *Methodus*, 5.183, 334; Bodin, *Method*, p. 146. Latin: 'tanta consuetudinis ac disciplinae vis est in rebus naturalibus et humanis, ut paulatim abeat in mores, et naturae vim obtineat'.
42 Jouanna, *Greek Medicine from Hippocrates to Galen*, Chapters 8 and 9; Mikkeli, *Hygiene in the Early Modern Medical Tradition*; Van Hoof, 'Plutarch's "Diet-Ethics"'.

In the Renaissance, the field of dietetics or hygiene (the branch of medicine concerned with the preservation and restoration of health through a regulated manner of living) was still dominated by a Galenic template that postulated the existence of six 'non-natural' factors influencing human health: ambient air, food and drink, motion and rest, wake and sleep, excretion and retention, and the passions of the mind.[43] The correct administration of these six factors was deemed crucial for human wellbeing and formed the object of dietetics. Dietitians would typically advise their patients on what they should eat and how long they should sleep depending on the patient's gender, age, and individual constitution, but also taking into account other factors such as the season of the year or the nature of the local climate. Stressing the importance of 'good air' for the wellbeing of a person, they offered remedies for those who lived in places where the air was 'unduly warm or cold, dry or moist',[44] and therefore dangerous for human health. They explained, for instance, how careful regulation of food intake and sleep patterns could work as a remedy against unfavourable climatic conditions. Diet—in this broader sense of 'life regimen'—thus became a popular and relatively accessible way of coping with environmental influence, and such it remained throughout and beyond the Renaissance.[45]

The sixteenth and early seventeenth centuries witnessed the publication of a wealth of healthcare books reconnecting to this longstanding tradition: among other things, these works offered detailed advice on how to regulate one's diet and lifestyle in order to counterbalance the effects of air on one's temperament.[46] The framework adopted in these texts was still largely that of Galenic humoral theory, itself based in turn on ancient Greek elemental theory. The gist of it is simple: there exist four elements (fire, earth, air, water), four qualities (hot, cold, dry, wet), four humours (blood, phlegm, yellow bile or choler, black bile or melancholy), and four temperaments (sanguine, phlegmatic, choleric, melancholic); health consists in a good balance between these different components, and while slight excess (say, a slightly choleric or a slightly melancholic temperament) is perfectly normal

43 In general see Temkin, *Galenism*. More specifically on the persistence of Galenic dietetics in the Renaissance, see Siraisi, *Avicenna in Renaissance Italy* and Albala, *Eating Right in the Renaissance*.

44 Galen, *Hygiene*, 1.4, 11.

45 On the medieval tradition, see Mikkeli, *Hygiene in the Early Modern Medical Tradition*; Nicoud, *Les Régimes de santé au Moyen Âge*. For the early modern tradition, see Cavallo and Storey, *Healthy Living in Late Renaissance Italy*.

46 On the popularity of healthcare books in the early modern period, see Slack, 'Mirrors of Health and Treasures of Poor Men'.

in a healthy body, greater excesses can result in *diskrasia* (temperamental imbalance) and lead to serious illness. Behind this deceptive simplicity, however, lies great complexity. Temperamental theory was a rather loose and adaptive framework that could accommodate disagreements over points of detail: working within that structure, Renaissance physicians (but also thinkers without any formal medical training, such as Bodin) were able to develop highly idiosyncratic doctrines—sometimes as a result of different interpretations of the same authoritative texts—without, however, departing from the fundamental principles that were largely shared among their community.

The work of Nicolas Abraham de La Framboisière—court physician during the reign of Henry IV, and professor of medicine at Reims—is a good example of how consensus and disagreement could coexist in medical approaches to climate theory in the Renaissance. La Framboisière's case is in many ways representative of contemporary trends: for instance, his choice of devoting a whole book of his treatise on healthcare (*Le Gouvernement necessaire à chacun pour vivre longuement en santé*, 1600) to examining diet in relation to place and climate is shared by many other medical writers of the time.[47] The same can be said more generally of his ideas on the transformative powers of food (which, as we shall see, represent another interesting angle for thinking about the continuum between nature and culture in the Renaissance). Nevertheless, as soon as we move into a more detailed analysis of his ideas on climate, temperament, and diet, we start noticing significant discrepancies between La Framboisière's views and those of at least some of his contemporaries.

La Framboisière's *Gouvernement* is divided into eight books, each of which examines the question of regimen from a particular perspective. Having outlined general rules for the preservation of health (Book 1) and discussed how regimens must vary according to individual temperament (Book 2), gender (Book 3), and age (Book 4), in Book 5 La Framboisière turns to considering the relationship between regimen and place. After restating the classic Galenic view that climate shapes temperament, the royal physician calls attention to the necessity of following different lifestyles in different countries: 'il est besoin és regions chaudes de se gouverner autrement qu'és

47 This includes La Framboisière's older colleague Joseph Duchesne (also known as Quercetanus, 1544–1609), who was also active at the court of Henry IV for a brief period at the closing of the sixteenth century. His treatise on healthcare, available in both French (*Le Pourtraict de la santé*, Paris: Claude Morel, 1606) and Latin (*Diaeteticon polyhistoricon*, Paris: Claude Morel, 1606), contained several sections on the relationship between food and climatic influence.

regions froides, et aux autres lieux humides autrement qu'aux lieux secs' ('one must govern oneself differently in hot regions than in cold regions, and in humid places than in dry places').[48] As he proceeds to offer some concrete examples of how such a climate-specific health regimen works in practice, La Framboisière takes a rather controversial stance by associating western peoples with a phlegmatic (i.e. cold and humid) temperament and eastern peoples with a choleric (i.e. dry and hot) temperament. This view contrasted with longstanding ethnic stereotypes that portrayed Asian peoples as soft and effeminate as opposed to the strong and manly European peoples, which we find, for instance, in the Hippocratic treatise *Of Airs, Waters, and Places* (fifth century BC).[49] Inversely, for La Framboisière, easterners are 'harder, manlier, braver, and more courageous' than westerners on account of their choleric constitution. While this may seem like a positive assessment of their nature, La Framboisière immediately adds that the excess of bile in their body makes them subject to a range of hot diseases, which can only be avoided through an appropriate corrective regimen:

> Les Orientaux sont subiects aux maladies chaudes, provenantes d'humeurs bilieuses. Partant ont besoin d'user de viandes raffraichissantes, et de mettre force eau en leur vin, et feront mieux de vendre leurs espiceries aux autres nations, que de s'en servir. Les bains d'eau douce leur sont proffitables. L'exercice violent, la cholere, et toutes autres choses qui eschauffent et dessechent fort, leur sont nuisibles. Le dormir leur est bon, et le coïte souvent contraire.[50]

> ([Easterners] need to eat cooling foods and to put abundant water in their wine, and they would do better to sell their spices to other nations than consume them. Sweet-water baths are beneficial to them. Intense exercise, anger, and everything else that has a strong warming and drying effect are harmful to them. Sleep is good for them, and frequent intercourse bad).

48 La Framboisière, *Gouvernement*, p. 300.

49 Ibid., pp. 303–305. It is slightly paradoxical that La Framboisière should reach this anti-Hippocratic conclusion precisely through an excess of Hippocratism: his identification of western peoples as choleric and of eastern peoples as phlegmatic derives from the fact that, following Hippocrates against Aristotle, La Framboisière establishes a direct correlation between nature of the climate and nature of the people living in it ('puisque les personnes tiennent tousiours de la nature de leur pays, il ne faut point douter que les Orientaux ne soyent chauds et secs', p. 303). Because he considers eastern climates to be drier and hotter than western climates due to the presence of the rising sun, he must also conclude that people living in the east have drier and hotter temperaments than people living in the west.

50 Ibid., p. 304.

La Framboisière gives similar—but opposite—advice to the phlegmatic western peoples, who should instead 'user d'une manière de vivre chaude et sèche' ('follow a hot and dry regimen') to compensate for their naturally cold and humid constitution: this includes long walks, intense physical exercise, and liberal consumption of strong wine, roasted meats (as opposed to boiled meats), spices and other hot condiments; on the other hand, foods such as fish, soups, fruits and salads should be avoided, and sexual appetites carefully managed.[51]

La Framboisière's prescriptions for southerners follow a similar course of reasoning, on the assumption that their nature should be hot like that of the climate in which they live.[52] But when it comes to northerners, La Framboisière surprisingly breaks the pattern. He claims that, of all people, northerners alone possess a temperament that is not an exact mirror of their country: though cold on the outside, they are hot and humid on the inside 'due to the coldness of the region, which prevents the dissipation of spirits'.[53] This unexpected application of the Aristotelian (and later Galenic) principle of antiperistasis, in a discussion of climatic influence largely inspired by Hippocrates, is a good example of how the interplay of different sources could generate unusual and sometimes puzzling results within the context of an apparently static and repetitive tradition. It is also an indication of how difficult it was for Renaissance authors to challenge especially powerful ethnic stereotypes such as that of the sanguine, strong-bodied, resourceful northerner, which itself rested on the premise of a hot and humid northern constitution. While La Framboisière, as we have seen, does not hesitate to overturn other longstanding ethnic stereotypes in his work, the exception that he makes in the case of northern peoples might relate to the special place that the north occupies in his personal map of the world, which positions Europe in the northern quadrant and thus identifies France as a northern region (albeit the most temperate of all).[54]

51 Ibid., p. 305.

52 Ibid., pp. 306–308. Southerners must adopt a regimen close to that of easterners, though stricter than the latter on account of their hotter nature: thus consumption of wine is altogether discouraged, while a special word is said in favour of seasoning food with the juice of oranges, lemons, and pomegranates.

53 Ibid., p. 309 ('Partant les Septentrionaux bien qu'ils tiennent de la complexion du pays, si ont ils neantmoins dans le corps abondance de chaleur naturelle et d'humeur radicale, à cause de la froidure de la region, qui empesche la dissipation des esprits'). There is an explicit mention of Galen's *De regimine sanitatis* a few lines above this passage.

54 Ibid., p. 311 ('Toute la terre est divisée en quatre parties, l'Europe située du costé de Septentrion, l'Asie au levant, l'Afrique au Midy, et l'Amerique vers le Ponant. L'Europe [...] contient-elle [...] beaucoup de regions, dont la Gaule est la plus temperée de toutes [...] d'autant qu'elle est

As this brief analysis has made clear, La Framboisière's discussion of climatic influence and corrective regimen features a heady mixture of traditional and unconventional ideas. While La Framboisière's views differ from those of his contemporaries in several important respects, the fundamental insight that inspires Book 5 of his *Gouvernement*—that the influence of ambient air can be counterbalanced through careful regulation of a person's diet and lifestyle—is one to which most, if not all, medical writers in the Renaissance readily subscribed.[55] In particular, his thoughts on the transformative effects of food belong in a much wider culture of thinking about the relationship between nature and nurture in dynamic and non-deterministic terms. From a humoral perspective, food itself is, in a sense, nature-turned-culture, as humans appropriate the fruits of the earth (and the nonhuman animals that live on it) not only to *sustain* themselves (a natural need that can be satisfied through instinct), but to *transform* themselves through autonomous acts of self-fashioning (a cultural gesture that connects the sphere of dietetics to that of ethics). The doctrine of corrective regimen highlights this cultural dimension of eating not only by calling for greater awareness when choosing one's diet, but also by stressing that the natural properties of foods can be artificially modified through different cooking methods (e.g. grilling v. boiling) or the use of certain condiments (e.g. spices or lemon juice).[56] Furthermore, this food which is nature-turned-culture is itself reconverted into nature as soon as it is eaten, assimilated, and transformed into humours, flesh, and vital heat. Corrective regimens, as described in the works of La Framboisière and of innumerable medical writers from the Renaissance, are ultimately nothing else but the art of governing this open-ended relationship between nature and culture, with a view to turning human nature into a 'second nature' which is neither entirely nature nor entirely culture, but rather which inhabits the space in-between the two.

iustement située au milieu des quatres pays notables qui l'environnent de tous costez', namely Italy, Spain, England, and Germany). Stereotypes relating to the north/south divide were in any case more powerful than those relating to the east/west divide; for instance, Bodin explicitly states that the former distinction is more relevant than the latter one (Bodin, *Les Six Livres de la République*, 5.1, 690).

55 For a rare counterexample, one can see Huarte's *Examen de ingenios para las ciencias*, which propounds a much less optimistic view of the powers of regimen.

56 See Albala, *Eating Right*, for other examples.

Conclusions

This essay has offered a reappraisal of Renaissance climate theories that shifts attention from notions of causality and determinism to notions of embeddedness, embodiment, and dynamic correlation, in order to highlight what Jean-Baptiste Fressoz has recently called the environmental 'reflexivity' of modern societies.[57] In social theory, 'reflexivity' designates a bidirectional relationship between causes and effects that mutually influence each other in an open-ended, spiralling process. As Jean-Patrice Courtois and Catherine Larrère have shown for eighteenth-century France—and as this chapter has sought to demonstrate for an earlier period—climate theory is inherently reflexive in this specific sense, because it calls attention to the manifold 'transactions' between humans and the physical 'environments' in which they live.[58] From this perspective, climate theory consists not in *establishing* but in *abolishing* any rigid dualism between nature and culture, and in thinking about the particular 'epistemic space' that is thus opened between these two dimensions.[59] The three authors discussed in this essay all testify, from their own unique perspectives, to this dynamic, non-dualistic, and anti-deterministic spirit of early modern climate theory. Whether it is by reintegrating man within a cosmic process of vicissitude that binds the human and the natural together (Le Roy); by proposing a 'second nature' born of continued discipline against the influence of climate (Bodin); or by examining the transformative powers of food as part of an open-ended relationship between nature and culture (La Framboisière), each author considered here frames the relationship between humans and their living environments in transactional terms—as a complex set of mutual interactions, negotiations, and exchanges that constantly redefine the very subjects involved in this relationship. Contrasting strongly with the dualism and essentialism implicit in the modern idea of a clear divide between nature and culture, climate theories may thus provide us with a helpful starting point for rethinking our connection to the natural world in the form of an 'integrated ecology' of human-nature relationships.[60]

57 Fressoz, *L'Apocalypse joyeuse*, p. 13.

58 Courtois, 'The Climate of the *Philosophes*'; Courtois, 'Le Physique et le moral'; Larrère, 'Montesquieu et l'espace'.

59 Fressoz, *L'Apocalypse joyeuse*, p. 13. Georges Benrekassa has similarly spoken of the 'common space' (*espace commun*) that climate theory opens up between man and nature (*La Politique et sa mémoire*, p. 207).

60 Latour, *We Have Never Been Modern*; Descola, *Beyond Nature and Culture*.

Works Cited

Albala, Ken, *Eating Right in the Renaissance* (Berkeley, Los Angeles and London: University of California Press, 2002).

Benrekassa, Georges, *La Politique et sa mémoire: le politique et l'historique dans la pensée des Lumières* (Paris: Payot, 1983).

Bodin, Jean, *Method for the Easy Comprehension of History*, trans. by Beatrice Reynolds (New York: Columbia University Press, 1945).

———, *Methodus ad facilem historiarum cognitionem*, ed. by Sara Miglietti (Pisa: Edizioni della Normale, 2013).

———, *Les Six livres de la République* (Paris: Jacques Dupuys, 1579).

Cavallo, Sandra, and Tessa Storey, *Healthy Living in Late Renaissance Italy* (Oxford: Oxford University Press, 2013).

Courtois, Jean-Patrice, 'The Climate of the *Philosophes* during the Enlightenment' in *Climates Past and Present: Perspectives from Early Modern France*, ed. by Sara Miglietti, special issue of *Modern Language Notes* 132 (September 2017).

———, 'Le Physique et le moral dans la théorie du climat chez Montesquieu', in *Le Travail des Lumières. Pour Georges Benrekassa*, ed. by Caroline Jacot Grapa et al. (Paris: Champion, 2002), pp. 139–156.

Couzinet, Marie-Dominique, *Histoire et méthode à la Renaissance: une lecture de la Methodus ad facilem historiarum cognitionem de Jean Bodin* (Paris: Vrin, 1996).

Descola, Philippe, *The Ecology of Others*, trans. by Geneviève Godbout and Benjamin P. Luley (Chicago: Prickly Paradigm Press, 2013).

———, *Beyond Nature and Culture*, trans. by Janet Lloyd (Chicago: University of Chicago Press, 2013).

Duchesne, Joseph, *Le Pourtraict de la santé* (Paris: Claude Morel, 1606).

———, *Diaeteticon polyhistoricon* (Paris: Claude Morel, 1606).

Floyd-Wilson, Mary, *English Ethnicity and Race in Early Modern Drama* (New York: Cambridge University Press, 2003).

———, and Garrett Sullivan, 'Introduction: Inhabiting the Body, Inhabiting the World', in *Environment and Embodiment in Early Modern England*, ed. by Mary Floyd-Wilson and Garrett Sullivan (New York: Palgrave Macmillan, 2007), pp. 1–13.

Fressoz, Jean-Baptiste, *L'Apocalypse joyeuse: une histoire du risque technologique* (Paris: Seuil, 2012).

Galen, *A Translation of Galen's Hygiene. De sanitate tuenda*, trans. by R. Montraville Green (Springfield: Charles C. Thomas, 1951).

Gingerich, Owen, 'Sacrobosco as a Textbook', *Journal for the History of Astonomy* 19 (1988), 269–273.

Glacken, Clarence, *Traces on the Rhodian Shore: Nature and Culture in Western Thought from Ancient Times to the End of the Eighteenth Century* (Berkeley, Los Angeles and London: University of California Press, 1967).

Huarte, Juan, *Examen de ingenios para las ciencias* (Baeza: Juan Bautista de Montoya, 1575).

Hulme, Mike, 'Reducing the Future to Climate: A Story of Climate Determinism and Reductionism', in *Klima*, ed. by James R. Fleming and Vladimir Jankovic, special issue of *Osiris* 26.1 (2011), 245–266.

Hunt, John Dixon, *Greater Perfections: The Practice of Garden Theory* (London: Thames & Hudson, 2000).

Jouanna, Jacques, *Greek Medicine from Hippocrates to Galen: Selected Papers* (Leiden: Brill, 2012).

Larrère, Catherine, 'Montesquieu et l'espace', in *Espace et lieu dans la pensée occidentale: de Platon à Nietzsche*, ed. by Thierry Paquot and Chris Younès (Paris: La Découverte, 2012), pp. 147–169.

Latour, Bruno, *We Have Never Been Modern,* trans. by Catherine Porter (Cambridge, Mass.: Harvard University Press, 1997).

Le Roy, Louis, *De la vicissitude ou varieté des choses en l'univers* (Paris: Pierre l'Huilier, 1575).

———, *Of the Interchangeable Course, or Variety of Things in the Whole World,* trans. by Robert Ashley (London: Charles Yetsweirt Esq., 1594).

Lestringant, Frank, 'Europe et théorie des climats dans la seconde moitié du XVIe siècle', *La Conscience européenne au XVe et au XVIe siècle* (Paris: Editions de l'ENSJF, 1982), pp. 206–226.

Lovejoy, Arthur O., *The Great Chain of Being* (Cambridge, MA.: Harvard University Press, 1936).

Miglietti, Sara, and John Morgan, 'Introduction: Ruling "Climates" in the Early Modern World' in *Governing the Environment in the Early Modern World: Theory and Practice*, ed. by Sara Miglietti and John Morgan (London and New York: Routledge, 2017), pp. 1–21.

———, 'New Worlds, Ancient Theories: Reshaping Climate Theory in the Early Colonial Atlantic' in *Translating Nature: A Transcultural History of Early Modern Science in the Atlantic World*, ed. by Jaime Marroquín Arredondo and Ralph Bauer (Philadelphia: University of Pennsylvania Press, forthcoming).

Mikkeli, Heikki, *Hygiene in the Early Modern Medical Tradition* (Helsinki: Academia Scientiarum Finnica, 1999).

Nicoud, Marylin, *Les Régimes de santé au Moyen Âge: naissance et diffusion d'une écriture médicale en Italie et en France (XIIIe-XVe siècle)*, 2 vols. (Rome: École Française de Rome, 2007).

Pinna, Mario, *La teoria dei climi: una falsa dottrina che non muta da Ippocrate a Hegel* (Rome: Società geografica italiana, 1988).

Serres, Michel, *The Natural Contract,* trans. by Elizabeth MacArthur and William Paulson (Ann Arbor: The University of Michigan Press, 1995).

Severini, Maria Elena, 'La vicissitudine o mutabile varietà delle cose. La traduzione italiana di Ercole Cato' in *Loys Le Roy, renaissance et vicissitude du monde*, ed. by Danièle Duport (Caen: Presses Universitaires de Caen, 2011), pp. 139–152.

Siraisi, Nancy, *Avicenna in Renaissance Italy: The Canon and Medical Teaching in Italian* (Princeton: Princeton University Press, 1987).

Slack, Paul, 'Mirrors of Health and Treasures of Poor Men: The Uses of the Vernacular Medical Literature of Tudor England' in *Health, Medicine and Mortality in the Sixteenth Century*, ed. by Charles Webster (Cambridge: Cambridge University Press, 1979), pp. 237–273.

Spavin, Richard, 'Jean Bodin and the Idea of Anachorism' in *Governing the Environment*, ed. by Sara Miglietti and John Morgan, pp. 38–55.

Staszak, Jean-François, 'Nature et culture: des origines du "déterminisme géographique"', *Géographie et cultures* 19 (1996), 95–115.

———, and Marie-Dominique Couzinet, 'À quoi sert la "théorie des climats"? Éléments d'une histoire du déterminisme environnemental', in *Géographie et philosophies*, ed. by Marie-Dominique Couzinet and Marc Crépon, special issue of *Corpus. Revue de philosophie* 34 (1998), 9–43.

Suggi, Andrea, 'Cronologia e storia universale nella Methodus di Jean Bodin', *I castelli di Yale. Quaderni di filosofia* 3 (1998), 75–92.

Temkin, Oswei, *Galenism: The Rise and Decline of a Medical Philosophy* (Ithaca, NY, and London: Cornell University Press, 1973).

Thorndike, Lynn, *The Sphere of Sacrobosco and Its Commentators* (Chicago: Chicago University Press, 1949).

Valleriani, Matteo, 'The Tracts on the "Sphere": Knowledge Restructured over a Network', in *The Structures of Practical Knowledge*, ed. by Matteo Valleriani (Cham: Springer, 2017), pp. 421–473.

Van Hoof, Lieve, 'Plutarch's "Diet-Ethics": Precepts of Healthcare between Diet and Ethics', in *Virtues of the People: Aspects of Plutarchean Ethics*, ed. by Geert Roskam and Luc Van der Stockt (Leuven: Leuven University Press, 2011), pp. 109–129.

Weisinger, Herbert, 'Louis Le Roy on Science and Progress', *Osiris* 11 (1954), 199–210.

Williams, Raymond, *Keywords: A Vocabulary of Culture and Society* (Oxford: Oxford University Press, 1985).

Zacharasiewicz, Waldemar, *Die Klimatheorie in der englischen Literatur und Literaturkritik von der Mitte des 16. bis zum frühen 18. Jahrhundert* (Vienna: W. Braumüller, 1977).

About the Author

Sara Miglietti (Senior Lecturer in Cultural and Intellectual History, The Warburg Institute, University of London) has a PhD in philosophy jointly awarded by the Scuola Normale Superiore of Pisa and the Université Paris Descartes (co-tutelle de thèse, 2012). Prior to joining the Hopkins faculty in 2015 she was a lecturer at the University of Warwick (2013–2015) and at Georgetown University (2012–2013), and a research fellow at the University of Fribourg, Switzerland (2012–2014). She has been the recipient of visiting fellowships at various institutions, including the École Normale Supérieure in Paris, the École Nationale des Chartes, the École Normale Supérieure in Lyon, Harvard University, the Institut d'Histoire de la Réformation in Geneva, the Warburg Institute, and the Institute for Historical Studies in London. Her current project, Mastering the Climate: Theories of Environmental Influence in the Long Seventeenth Century, investigates the dissemination of 'climate theories' across early modern Europe and the colonial world. She is also co-editing a bilingual (Latin/French) edition of Jean Bodin's *Six livres de la République* (with Mario Turchetti), as well as two collections of essays: *Reading Publics in Renaissance Europe, 1450–1650* (with Sarah E. Parker) and *Ruling Climate: The Theory and Practice of Environmental Governmentality, 1500–1800* (with John Morgan).

7. Almost Encountering Ronsard's Rose

Phillip John Usher

Abstract
This chapter takes up the French poet's most famous ode 'Mignonne, allons voir si la rose…' in order to ask a simple but important question: what are the barriers to close-reading a poem such as this one, a poem made of 'signs', if we (also) try to access through it the nature—or Nature—of which it perhaps claims to be an imitation? To explore such a question, Usher experiments with three ways of reading the ode. He first explores the cultural/historical approach offered by book history. A second approach seeks out connections between Ronsard's poem and early modern botany's own discussion of roses. The third and final method strives to get beyond the poem as cultural artefact by drawing on contemporary plant theory (Jeffrey Nealon, Michael Marder, Luce Irigaray).

Keywords: Pierre de Ronsard, rose, nature, ode, botany, plant theory

The most famous poem of early modern France—perhaps of all French literature—is a poem about a plant.[1] And yet the combined forces of anthropocentrism, zoocentrism, and historicism have made it very difficult to perceive that plant *as plant*, trapping the poem and its readers, across the centuries, in the purified domain of the cultural. The poem in question, of course, is Ronsard's 'Mignonne, allons voir si la rose…' ('Beloved, let us go see if the rose…'), which first appeared in a sort of

1 Some of the ideas in this chapter were presented in keynotes given at a recent conference titled 'Parenthetic Modernity' at Linköping University, Sweden, and at the 'Joy of Close Reading' conference organized at Syracuse University, USA, in honor of the late Hope Glidden. I should like to thank Carin Franzén and Jesper Olsson, and Albrecht Diem and Stephanie Shirilan for their respective invitations, and fellow speakers and audience members for their productive questions and useful feedback.

Goul, P. and P.J. Usher (eds.), *Early Modern Écologies. Beyond English Ecocriticism.*
Taylor & Francis Group, 2020
DOI 10.5117/9789462985971_CH07

appendix to the 1553 edition of the poet's *Amours*, one of what the volume's title refers to as 'quelques Odes de L'auteur, non encor imprimées' ('a few odes by the author, not previously printed').[2] In these verses, the poet and his beloved head out to look at a rose that had been in full bloom that very morning, only to discover that its petals have fallen to the ground over the course of just one day. In the third stanza, the poet concludes by offering up a lesson *not* about the rose or about plant life, but about human mortality: 'cueillés, vôtre jeunesse' ('gather the bloom of your youth'), a version of the *carpe diem* motif that is omnipresent in Ronsard's writings.[3] The poem clearly *is*, as we have all been taught, about the passing of youth, about seizing the day, and about human joy and sadness—but need that necessarily lead us to ignore the rose *as rose*? Might we not ask: what of the plant itself? Must our cultural readings delete it?

This is, for sure, far from the only poem in which Ronsard features a rose. The word appears a total of 264 times throughout his collected works.[4] But it is without a doubt this poem that inspires critics refer to Ronsard not only as the 'prince of poets' and the 'poet of princes', but also as the 'poet of roses', and it is thus the best place to open the present reflection.[5] The cultural hold on the poem is powerful: almost all commentary on the poem foregrounds the *carpe diem* motif to the exclusion of the rose as rose. The point barely needs a footnote, but a useful and representative flashpoint can be found in the entry for 'Fleurs' (Flowers) in the *Dictionnaire de Pierre de Ronsard*, which emphases how, in Ronsard, flowers are turned into metaphors and symbols or otherwise mythologized, most frequently to sketch out a comparison between flowers and (female) beauty.[6] Dominique Brancher, in an otherwise compelling book about libertine botany, paints

2 Ronsard, *Les Amours de P. de Ronsard Vandomois*. On this edition, see Barbier, *Ma Bibliothèque poétique*, pp. 36–43 (items 10–11). The publication of 'Mignonne' was far from the only important literary event in 1553, although it has come to overshadow everything else, a situation explored in a conference at the *Bibliothèque nationale de France* in 2008, 'Paris, 1553. Audaces et innovations poétiques' (Paris, 1553. Poetic Boldness and Innovations). For a summary of the conference's main points, see Halévy and Vignes, 'Paris, 1553'.

3 On Ronsard's extensive deployment of this *topos*, see especially Yandell, *Carpe corpus*.

4 I base this figure on Creore, *A Word-Index to the Poetic Works of Ronsard*.

5 On this reputation as the 'poet of roses' see Dulmet, 'Ronsard, poète des roses, des femmes et des princes'; Lafont, 'Rose, femme, événement: parcours d'un poncif'; and Livet, 'Sur le rosier: Ronsard fleurit la France'.

6 Rouget, *Dictionnaire de Pierre de Ronsard*, pp. 250–252. Along similar lines, see Simonin, '"Poésie est un pré", "Poème est une fleur": métaphore horticole et imaginaire du texte à la Renaissance' and Duport, *Les Jardins qui sentent le sauvage. Ronsard et la poétique du paysage*.

a similar picture, stoutly affirming that 'd'Érasme à Cyrano, la plante sert toujours un discours d'homme' ('from Erasmus right up to Cyrano de Bergerac, the plant always serves some human discourse').[7] How did such a situation come about? Why it is all but impossible to perceive a rose in this poem? These are the simple questions with which I begin. To unpack them, with an eye fixed on the wider nature–culture debates to which the Anthropocene forces us to respond, the reflection that follows is situated at the crossroads of opposing modes of reading—namely at the intersection of historicism and the nascent field of plant theory—in order to see how the tension between the two modes can ultimately enrich and nuance both.[8]

Although the poem is very well known, it is important first that we reread it:

Mignonne, allons voir si la rose
Qui ce matin avoit declose
Sa robe de pourpre au soleil,
A point perdu, cette vesprée,
Les plis de sa robe pourprée,
Et son teint au vostre pareil.

Las, voiés comme en peu d'espace,
Mignonne, elle a dessus la place
Las, las, ses beautés laissé cheoir!
O vraiment maratre Nature,
Puis qu'une telle fleur ne dure
Que du matin jusques au soir.

Donc, si vous me croiés, mignonne:
Tandis que vôtre âge fleuronne
En sa plus verte nouveauté,
Cueillés, cueillés, vôtre jeunesse
Comme à cette fleur, la vieillesse
Fera ternir vôtre beauté.[9]

7 Brancher, *Quand l'esprit vient aux plantes*, p. 104.

8 The bibliography for the nature-culture debates grows daily. See essentially Latour, *Nous n'avons jamais été modernes*; Haraway, *Companion Species Manifesto*; and Descola, *Par-delà nature et culture*.

9 Ronsard, *Œuvres complètes*, 5. 196–197.

> (Beloved, let us go see if the rose, which this morning had unfurled her crimson gown to the Sun, has not lost this evening the folds of her crimson gown and her complexion that resembles your own.
> Alas! See how in a short space of time, beloved, she has shed around her on the ground, alas, alas! her beauteous charms. O Nature, you are a truly unnatural mother, since such a flower lives only from morning until evening!
> So, if you will trust me, beloved, while your age is blossoming in its most verdant freshness, gather, gather the bloom of your youth; just as it does to this flower, old age will blight your beauty).[10]

It would be possible to write a long and very interesting study of the cultural history of the ascendancy or canonization of Ronsard's ode. Such a study would trace the progressive layering that made and still make the ode *that poem*. It would discuss how those verses were quickly set to music in *Le recueil des plus belles et excellentes chansons* [...] *tirées de divers autheurs* ('Compendium of the Most Beautiful and Excellent Songs [...] drawn from Various Authors', 1576).[11] It would examine how, very early on, the poem became a metonym for Ronsard's poetic output as a whole: in *La Cresme des bons vers* ('A Crop of Good Verses', 1622), the poem is featured at the very beginning of the florilegium's Ronsard section.[12] It would likely posit that, in addition to becoming quickly canonical, the poem swiftly received its canonical 'meaning', quoting such commentary as that found in the 1623 edition, which reads: 'La fleur et la jeunesse, sont de peu de durée, & leur usage encore a sa saison, laquelle il ne faut pas laisser perdre' ('Flower and youth do not last long—and they should be used in their correct season, which must not be allowed to pass').[13]

Such a study would have thus already established that by the early seventeenth-century, the poem's (cultural) status and its (cultural) meaning were largely in place. The trajectory could continue up until our own moment,

10 Ronsard, *Selected Poems*, ed. Quainton and Vinestock, p. 78. For an alternative (rhymed) translation see Shapiro (ed.), *Lyrics of the French Renaissance*, pp. 300–301.

11 Chardavoine, *Le recueil des plus belles et excellentes chansons en forme de voix de ville*, pp. 4–5. The setting to music of Ronsard's poetry has been studied *inter alia* in Thibault and Perceau, *Bibliographie des poésies de P. de Ronsard mises en musique au XVIe siècle* and in *Collarile, Ronsard et la mise en musique des* Amours *(1552–1553)*.

12 *La Cresme des bons vers, triez du meslange & cabinet des sieurs de Ronsard, dv Perron, de Malerbe, de Sigongnes, de Lingendes, Motin, Maynard, de Bellan, d'Vrfé, Theophile & autres*, pp. 181–182.

13 Ronsard, *Les Œuvres de Pierre de Ronsard* (1623), vol. 1, p. 384.

telling the story of the 1987 creation of the so-called 'Pierre de Ronsard rose', enumerating all the rose-derived products (soap, tea, rose-decorated porcelain) on sale at the gift shop at Sainte-Cosme, and commenting on screen shots of references to Ronsard's ode in *Pokémon Go*.[14] The cultural grasp on the poem is further strengthened by the fact that, as was known even by Ronsard's earliest readers, 'Mignonne' reworks a poem by the Latin poet Ausonius, 'De rosis nascentibus' ('On Budding Roses'). A glance at the closing verses confirms the proximity of the two poems: 'Conquerimur, Natura, brevis quod gratia talis' ('Nature, we grieve that such beauty is short-lived'); 'sed bene, quod paucis licet interitura diebus | succedens aevum prorogat ipsa suum. | collige, virgo, rosas, dum flos novus et nova pubes, | et memor esto aevum sic properare tuum' ('But 'tis well; for though in a few days the rose must die, she springs anew prolonging her own life. Then, maidens, gather roses, while blooms are fresh and youth is fresh, and be mindful that so your life-time hastes away').[15] When we read the ode, then, we can all too easily find ourselves caught up in our memory of the Latin text, and caught up in the reception history of the poem that makes it cultural artefact.[16] This is, of course, part of the story—but, again, what about the rose?

There are multiple ways in which a reader might inquire into the plantness of Ronsard's rose. From an historicist point of view, it is tempting to pay attention to the contemporaneity of Ronsard's ode and the development of early modern botany, an approach that (as far as I can tell) has not been attempted. Only one article is listed under the heading 'botany' in François Rouget's recent Ronsard bibliography—one which, moreover, deals with the poem 'La salade' ('The Salad'), as if the roses in the *Œuvres* do not in fact qualify for such treatment.[17] It is indeed surprising that the botanical context has been set aside; if the history of botany is long—winding back to Ancient Egypt and Greece, to founding figures such as Theophrastus, Empedocles, Aristotle, Anaxagoras, and Dioscorides—the sixteenth century

14 Ondra, *Taylor's Guide to Roses*, p. 215 and p. 390. The boutique is part of the Prieuré Saint-Cosme, rue Ronsard in La Riche, not far from the Université de Tours—see http://www.prieure-ronsard.fr/ (accessed 27 September 2019). I should like to thank Charles-Louis Morand-Métivier of the University of Vermont for alerting me to the presence of a reference to Ronsard's ode in *Pokémon Go*.

15 Ausonius, 'De rosis nascentibus' in Ausonius, *Works*, 2. 276–281, vv. 41 and 47–50.

16 On the intertextual ties that bind Ausonius and Ronsard, see Laumonnier, *Ronsard, poète lyrique*, p. 583 and Lafont, 'Rose, femme, événement'. See also Cuadraro, 'Ronsard en el arco tensado entre Ausonio y Guillen'.

17 Johnson, 'La salade tourangelle de Ronsard'.

witnesses a number of key evolutions. An iconic instant in this respect is Luca Ghini's foundation, eight years before the first publication of 'Mignonne, allon voir si la rose', of Europe's first botanical garden in Pisa (the *Orto Botanico*), which inaugurated in a very concrete way a shift from thinking about plants in terms of their properties and uses in human medicine, to recognizing their plant-ness.[18] But it is not just this one moment that counts: over the past couple of decades, scholars such as Paula Findlen, Brian Ogilvie, Sachiko Kusukawa, and Florike Egmond have shown the extent to which botany reinvents itself in the sixteenth century, for various reasons and with various consequences.[19] These authors show that *inter alia* the science of plants comes to be progressively less interested in the medical properties of plants and more in plants as plants, and in plants for plants' sake.

The first botanical garden in France would only appear at the end of the century when Henri IV established by *lettres patentes* Montpellier's *jardin des plantes* in 1593, to be directed by French botanist Richer de Belleval (i.e. almost a decade too late for Ronsard to have visited), but botanical sciences were nonetheless in full evolution in France earlier in the century.[20] To establish this, a few key names and dates will suffice. According to historians of botany, the French physician Jean Ruel, Ruelle, or Ruellius (1474–1537), a contemporary of Rabelais and predecessor or Ronsard, made a major intervention in thinking about plants and plant-ness. Although he was a physician and although in 1516 he published a Latin translation of Dioscorides's *De materia medica* (a pharmacopoeia that details the medicines that can be obtained from plants), in his own work, especially the *De Natura stirpium* (1536), Ruel asserted that 'botany is botany, and that pharmacy, like agriculture, pomology, and horticulture, is but one of its departments' and that all of them must remain 'subsidiary to the philosophy of plant life as a whole'.[21] The most famous botanical treatise of the time—written by

18 For a brief introduction to the history of Pisa's botanical garden, see Bedini, *L'Orto Botanico di Pisa: Piante, storia, personaggi, ruoli.*

19 Findlen, *Possessing Nature: Museums, Collecting, and Scientific Culture in Early Modern Italy*; Ogilvie, *The Science of Describing: Natural History in Renaissance Europe*; Kusukawa, Sachiko. *Picturing the Book of Nature: Image, Text, and Argument in Sixteenth-Century Human Anatomy and Medical Botany*; Egmond, *The World of Carolus Clusius: Natural History in the Making, 1550–1610* and *Eye for Detail: Images of Plants and Animals in Art and Science, 1500–1630*. See also the collective volume by Egmond, Hoftijer, and Visser, *Carolus Clusius: Toward a Cultural History of a Renaissance Naturalist.*

20 On the foundation and evolution of France's first botanical garden, Rossi, *Le Jardin des plantes de Montpellier: de la médecine à la botanique* and Rioux, *Le Jardin des plantes de Montpellier: quatre siècles d'histoire.*

21 Greene, *Landmarks of Botanical History*, p. 598.

Leonhart Fuch and also called *De Historia Stirpium* (1542)—was soon translated into French under the title *Commentaires tres excellens de l'hystoire des plantes* ('Most Excellent Commentaries on the History of Plants'), probably in 1549 and almost certainly before Ronsard published his famous ode.[22] Finally, in 1557, the French botanist Charles de l'Écluse (then at the start of his career) would translate Rembert Dodoens's *Cruydeboeck* under the title *Histoire des plantes* ('History of Plants').[23] In other words—and in historicist mode—there is certainly grounds for asserting that there is a clear historical overlap between the rise of botany in early modern France and the writing of Ronsard's famous ode, and potential to infer some kind of circulation (of ideas, of percepts, etc.) common to the two.

There is, however, no evidence (as far as I can tell) that Ronsard ever owned, consulted, or cared about Luca Ghini, Jean Ruel, Leonhart Fuchs, Rembert Dodoens, Charles de l'Écluse, or any other contemporary botanist. There is some reason to think that he *might* have read them, and that those texts *might* have had a direct impact on his poetry—Ronsard did, after all, draw on a whole host of non-literary texts while writing poetry. His 1560 edition of Gerolamo Cardano's *De subtilitate libri XX,* in which he underlined threes lines about the notion of fire ('Ignem/flamma')—*fire is burning air*—shaped the poet's conception of the mechanics and effect of the flame of love;[24] his 1530 copy of a Greek poem on the nature of venomous snakes, Nicander of Colophon's *Theriaca*, informed several sonnets about the poison of love;[25] and it has been shown that in his poetry he drew heavily on his 1558 editions of the *Works* of Hippocrates, especially regarding the symptoms and causes of disequilibria caused by humoral imbalances, fevers, melancholy, *hydropsie* (i.e. edema), coughing, and other illnesses, all of which leave their mark in Ronsard's poems. But none of the countless articles and studies about Ronsard's library, books, or reading habits mention works of early modern botany.[26]

Such historicist inquiries already suggest that there might be more to 'Mignonne' than the *carpe diem* motif, but they can only take us so far. In order to suggest another way of shifting the nature–culture balance in our

22 Fuchs, *Commentaires tres excellens de l'hystoire des plantes*.

23 Dodoens, *Histoire des plantes*. Charles de l'Écluse (aka Carolus Clusius) has received much attention of late, especially in Egmond, *The World of Carolus Clusius*.

24 Rouget, *Ronsard et le livre*, 1, pp. 58–59.

25 Ibid., 1, pp. 67–68.

26 Laumonier, 'Sur la bibliothèque de Ronsard'; Livingston, 'Notes sur la bibliothèque de Ronsard'; Labaste, 'Un nouveau livre de la bibliothèque de Ronsard'; Veyrin-Forrer, 'La bibliothèque de Ronsard'; Rouget, *Ronsard et le livre*.

reading of the poem, which will remain open to reading poetry alongside early modern botany without assuming direct connections, I should like to take a few steps back from the immediate context to situate the history of reading Ronsard's poem within a longer history of exclusion of plants within Western thought and metaphysics. To do this I turn, then, to the recent work of Emanuele Coccia, Matthew Hall, Luce Irigaray, Michael Marder, and Jeffrey Nealon, which I gather—borrowing the title of Nealon's book—under the general rubric of 'plant theory'.[27] Such works open a collective reflection about the reality of plant/vegetal life, in particular reacting (directly or indirectly) to what Nealon calls the 'foundational abjection of plant life' in Animal Studies', which he accuses of 'kingdomism'.[28] As much as our readings of Ronsard's ode have a history, so too does the exclusion of plants from Western metaphysics, which Hall analyses in the first chapter of his *Plants as People: A Philosophical Botany*, 'The Roots of Disregard'. Hall shows that there is no *originary* exclusion of plants, but rather an exclusion-*in-the-making*, especially after Plato, in whose writings can be detected 'a turning away from plants being viewed as related, active, autonomic beings'.[29] The exclusion of plants is here akin to that of women and slaves in Plato's thought. The exclusion is not total, to be sure: in the *Timaeus*, Plato indeed says that 'everything that partakes of life may be truly called a living being', but only before establishing a zoocentric hierarchy that sets plants apart for their lack of 'opinion or reason or mind' and before summing up by saying that plant nature is 'always in a passive state'.[30] Aristotle, for his part, pursues this 'drive toward separation and discontinuity', with plants now firmly set off as a 'lower class of being'.[31] Aristotle's nested hierarchy (in the *De Anima* and in *Parts of Animals*) of soul functions—growth/reproduction, locomotion/perception, and intellect—and the corresponding three degrees of soul—the nutritive soul of plants, the sensitive soul of animals, and the rational soul of human beings—'extends the Platonic separation of plants and animals', even as—via the *nesting*—it also recognizes certain continuities.[32]

27 Coccia, *La Vie des plantes. Une métaphysique du mélange*; Hall, *Plants as People: A Philosophical Botany*; Irigaray and Marder, *Through Vegetal Being: Two Philosophical Perspectives*; Marder, *Plant-Thinking: A Philosophy of Vegetal Life* and *Philosopher's Plant: An Intellectual Herbarium*; and Nealon, *Plant Theory: Biopower and Vegetable Life*. This is a burgeoning field of intellectual inquiry; many other titles could be adduced.

28 Nealon, *Plant Theory: Biopower and Vegetable Life*, p. 12.

29 Hall, *Plants as People: A Philosophical Botany*, p. 19.

30 Ibid., p. 21.

31 Ibid., p. 22.

32 Ibid., p. 23.

In his *Plant-Thinking: A Philosophy of Vegetal Life* (2013), Michael Marder summarizes the problem in a slightly different way: 'If animals have suffered marginalization throughout the history of Western thought, then non-human, non-animal living beings, such as plants, have populated *the margin of the margin*, the zone of absolute obscurity undetectable on the radars of our conceptualities', such that the 'suppression of the most basic question regarding plants became the breeding ground for their ethical neglect'.[33] If, in such a situation, the task at hand thus becomes that of giving 'a new prominence to vegetal life', it is by attending to the simplest of questions: 'How is it possible for us to encounter plants? And how can we maintain and nurture, without fetishizing it, their otherness in the course of this encounter?'[34] Such questions as these might, perhaps, help us read Ronsard's ode with fresh eyes. Responding to this challenge via concepts, nomenclature, and classification risks, in Marder's words, 'violating the flower' via a 'cognitive plucking' that leaves us only with a plant 'already dead and dry'.[35] Such is the 'Ronsard's rose' that the history of Ronsardian criticism hands us: a rose that is all cultural, that is symbol and symbol alone. In opposition to nominalism, conceptualism, and other cultural deadenings, Marder advocates recourse to 'hermeneutic phenomenology, deconstruction, non-Western thought, feminism, as well as to weak thought', the latter a reference to the *pensiero debole* of Gianni Vattimo.[36] Marder gathers these various resources because of what he calls their 'quasi-aesthetic receptivity' that can open up 'just enough space for the sunflower to grow without trimming it down to an object readily available for the subject's manipulation'—that is, to return to our present context, a method for allowing Ronsard's rose to be (also) *just a plant*.[37] Such a philosophical infrastructure as the one Marder proposes is weary of itself and chooses to be sympathetic to the methods of its object of study. The challenge, as Marder puts it, is 'to let plants be within the framework of what, from our standpoint, entails profound obscurity, which, throughout the history of Western philosophy, has been the marker of their life, [in other words] to allow plants to flourish on the edge or at the limits of phenomenality, of visibility and, in some sense of "the world"'.[38] Marder ultimately refers to this *debole* 'grasp' using the Portuguese word *desencontro*: an *encontro* that

33 Marder, *Plant-Thinking*, pp. 2–3.
34 Ibid., p. 3.
35 Ibid., p. 5.
36 Ibid., pp. 5–6. See Vattimo and Rovatti (ed.), *Il Pensiero debole*.
37 Ibid., p. 7.
38 Ibid., p. 9.

is not one, a *not-meeting*, a 'crossing of paths', since all 'we can hope for is to brush upon the edges of [the] being [of plants]'.[39]

In his book *Philosopher's Plant*, Marder occasions such a *desencontro* by nudging us in the direction of the thought of Luce Irigaray, who (he writes) 'urges us to listen to the muted vegetal rhythms in our life and thought, where growth has been stunted by the prejudices of metaphysics and the arrhythmia of modern existence'.[40] To unpack Marder's point and to gather materials for our method, we can turn to Irigaray's *J'aime à toi* (*I Love To You*) and in particular to a chapter titled 'L'amour entre nous', which we might render as 'Love Between Us', or perhaps—less literally, but perhaps more fully—as 'That Love We Share'.[41] There, we find an opposition between two manners of perceiving: one, which she calls (admittedly a little simplistically) 'Western', is appropriative, conceptually bound, intent on closure; the other, in which bodies are not a given but part of an ongoing sense of *being with*, posits a relationship to the world exemplified by the way that Buddha looks at a flower 'sans la cueillir' ('without plucking it'). In Irigary's words: Buddha 'regarde l'autre que lui sans l'enlever à ses racines' ('looks at that which is other than himself without detaching it from its roots').[42] The human-plant *desencontro* is a communion, a *vivre-avec* or *living-with*; it is about looking at plant being *not* to learn something (such as the shortness of human life), *not* to compare and contrast (my youth, too, passes like that of the flower), but in order to simply *be with*. Irigaray pursues this line of thought further in a book she subsequently co-wrote with Marder, *Through Vegetal Being*, which sketches out a manner of arriving at a dialogue across difference that *passes through the vegetal world*. All depends on how we *see*, about how we encounter the plant without making it merely a named something in our own world. Instead of a world that 'looks like a sort of museum composed of inanimate things invested with our projections' (such as Ronsard's poems have perhaps become), Irigaray wants to 'pass through the vegetal' to perceive a being-with-the-living.[43] Central to this is how she herself never says that Buddha looks at a lotus flower—she only ever says *flower*.

What happens if, after Marder and Irigaray, we relax our grasp on Ronsard's ode? We need not jettison the *carpe diem* motif completely, but we can set it aside temporarily, so that it might in fact return with greater

39 Ibid., p. 13.

40 Marder, *Philosopher's Plant*, p. 217.

41 Irigaray, *J'aime à toi*. All translations are mine.

42 Ibid., p. 49.

43 Irigaray and Marder, *Through Vegetal Being*, p. 85.

poignancy. What if we *try* to pay attention, 'in the present, to [the rose's] concrete singularity and [to] its sensible qualities, without substituting a name for them'?[44] Questions—simple questions—quickly arrive. Ronsard refers to the rose's 'robe de pourpre' and 'robe pourprée' (crimson gown). Are roses, in the early modern period, generally crimson? Does that matter? And, if we know the colour, can we ask: what kind of rose is this? We could even ask, would it make any difference if this were, say, an orchid or a daisy? Do those plants not also die quickly? What of the language of time here? How literally are we to think about a rose lasting from dawn to dusk? Are roses, more than, say, daffodils, particularly short-lived and hasty beings? Might there be in this respect an echo of early modern botanical sources and, if so, how might our reading of the poem change? And on the boundary of the *carpe diem/desencontro* readings, what of Ronsard's advice 'Cueillés, cueillés, vôtre jeunesse'—Ronsard may indeed encourage his *mignonne* to 'gather the bloom of her youth', but neither of them, in the poem, ever pluck the flower. Perhaps this is because it is too late... But the rose *is* still there at the end of the poem, still alive, albeit minus its petals. Does not this *remaining* matter? Let us pick up some of these threads, focusing first on plant colour, and secondly on plant time.

What, then, of the fact that, Ronsard's rose has a 'robe de pourpre'? Does that detail contribute to how the reader imagines a rose in its plant-ful singularity? How might the term have resonated with Ronsard's first readers? More or less specifically than today? A first point to note is that plant colour is something early modern botanists struggled with in several ways. Ancient authors, such as Pliny, had not paid much attention to the colour of plants; early modern botanists thus found themselves in the position of having to experiment to find a language capable of capturing different hues.[45] The situation, as one specialist has put it, was one of 'chromonymic chaos'.[46] When we read authors such as Leonhart Fuchs and Rembert Dodoens in their original languages or in their early modern French translations, we find all sorts of approximations and comparative paraphrases.

In Charles de L'Écluse's French version of Dodoens we find roses of different colours, described as follows: 'la Rose blanche' ('the white rose'); '[la rose] rouge [dont] les fleurs sont rouges' ('the red rose whose flowers

44 Ibid., p. 47.
45 '[L]es hommes de la Renaissance se trouvent en présence d'une multiplicité de plantes jusque-là inconnue, mais aussi en présence d'une multiplicité de couleurs nouvelles auxquelles les Anciens, comme Pline, n'avaient guère donné d'importance' (Selosse, 'Traduire les termes de couleurs', p. 1).
46 Ibid., p. 3. In French: 'un chaos chromonymique'.

are red'); the Provence rose, which is called 'vne espece moy[en]ne entre la Rose rouge & blanche' ('a type halfway between the red and white roses') whose flowers are 'ne rouges ne blanches' ('neither red nor white') but 'vne couleur moyenne entre le rouge et le blanc, bien pres incarnée' ('a colour midway between red and white, almost flesh coloured'); another kind whose flowers 'sont de belles couleur rouge obscur' ('of a fine dark red colour'); another that smells like cinnamon and which is 'de couleur palle en rouge' ('of a pale red colour'); another that is simply 'de couleur blanche' ('white in colour'); the wild rose that is 'de couleur blanche, ou tirant sur l'incarné' ('white in colour, or else close to flesh coloured'); and another 'de couleur blanche pour la pluspart, aucunefois rouge' ('normally white in colour, sometimes red').[47] All roses are thus presented as if on a colour continuum that stretches from white to red, with varying blends of both in between. Ronsard's rose with its 'robe de pourpre' would clearly find its place on this continuum, but it would be impossible to say from this which kind of rose it is. As the crossroads of historical verisimilitude and the impossibility of classification, we find ourselves—if we allow ourselves—in a *desencontro*.

Guillaume Guéroult's French rendering of Fuchs's text makes for an interesting comparison: there, we read that '[les roses] sont pareillem[en]t differ[en]tes en couleur, & en odeur' ('[roses] are equally different in colour and in smell')—but at no point does the text enumerate these different colours.[48] Most editions of Fuchs's work, in most languages, contain woodcuts, and in certain copies of the work these illustrations have been hand-coloured, for example those copies at the Bibliothèque nationale de France and at the US Agricultural Library. In the former of these, we see a rose plant of which some flowers are white, some pink, and some red—which, of course, means we are looking at an impossible plant! Once again, if we keep Ronsard's verses in mind, we *almost* encounter botany's rose. It is close by, but we escape the close ties of the kind that link 'Mignonne' to 'De rosis nascentibus'—and it is this escape that prepares the potential *desencontro*.

To variegate this *desencontro* and to get to its most important nexus, let us turn from plant colour to plant time. To get up close to the singularity of this rose, to move towards seeing *plant time* in this poem (rather than seeing plant time as a metaphor for human time), let us again draw on both theory and historicism: respectively, Marder's *Plant-Thinking* and early modern botany. The first section of Part II of Marder's book, 'Vegetal Existentiality',

47 Dodoens, *Histoire des plantes*, pp. 457–459.

48 Fuchs, *Commentaires tres excellens de l'hystoire des plantes*, f. Lvi^v.

focuses on 'the Time of Plants', in order to approach the question of plant being via the question of plant time, setting up a task 'to rethink temporality as the mainspring of the plants' ontology'.[49] Indeed, various plant processes relate to the passage of time, incluing germination, growth, flourishing, fermentation, decay, and dehiscence. More specifically, and drawing on Heidegger—namely the latter's 'hermeneutics of facticity' to explain the meaning of *Dasein*—Marder proposes that 'the meaning of vegetal being is time', which he subdivides into three categories: 1) 'the vegetal hetero-temporality of seasonal changes'; 2) 'the infinite temporality of growth'; 3) 'the cyclical temporality of iteration, repetition, and reproduction'.[50] To each of these, Marder dedicates a full section of Part II. Rather than follow Marder step-by-step here, I want to single out a couple of insights and build on them in a given direction. Most essential is his connection—after Heidegger, of course, but here for plants—of being and time. The most concrete way to grasp this is via the example of the hothouse in which humans can gain mastery 'over the time of plants' and thus 'manipulate their being'.[51] Such mastery does not involve negating some supposedly natural condition but, rather, interjecting into the plant a different time. Creating the conditions (more heat, less heat; more rain, less rain; more light, less light, etc.) that determine *when* a plant grows or flowers means mastery over plant being.[52] Because of this immediate connection between plant time and plant being, Marder notes that 'the plant's future is entirely contingent on alterity'—and that Other might be the climate of a given place, or indeed some agro-industrial complex.[53]

If, as Marder argues, vegetal temporality is wholly bound up with—not identical to, but impossible to separate from—plant being, then what is there, if anything, of plant time in Ronsard's ode? The end of Ronsard's poem 'Cueillés, cueillés, vôtre jeunesse' ('Gather, gather the bloom of your youth') translates, of course, Ausonius's 'college [...] rosas' ('gather your roses!')—and indeed, it gives voice to the *carpe diem* topos. The poem clearly evokes human time, as Ronsard does, poignantly, throughout his *Derniers vers* ('Last Verses'). That much we know. But the literary-historical fact of this traceable translation obfuscates the fact that, since the times of Theophrastus, plant time, even more than the description of shape or

49 Marder, *Plant-Thinking*, pp. 93–117; here p. 94.

50 Ibid., pp. 94–95.

51 Ibid., p. 102.

52 Ibid., p. 102.

53 Ibid., p. 107.

colour, has commonly been central to phytography. The emphasis on time in Ronsard's poem in many ways communes with the rose in a manner similar to that of both Theophrastus and his early modern inheritors. Book IV of Theophrastus's *Enquiry Into Plants* contains a long section that evokes plant temporality, notably both the 'comparative shortness of life of plants and trees' and how that temporality is different for different plants and between wild and domestic plants ('thus the wild olive pear and fig are longer-lived than the corresponding cultivated trees').[54] The lives of the apple tree and the pomegranate tree, for example, are singled out as particularly rushed, as are plants that grow too near to water, such as the white poplar or the elderberry tree.[55] Of particular interest is that the father of botany, in a manner that anticipates Marder, passes quickly from the question of *time* in a strict sense to that of *being* more generally—for instance when he muses on how trees change over their life-cycle: 'some trees, though they grow old and decay quickly, shoot up again from the same stock, as bay apple pomegranate and most of the water-loving trees [...] about these one might enquire whether one should call the new growth the same tree or a new one'.[56]

The centrality of time to plant life is asserted even more clearly in early modern botany: in Fuch's *Historia stirpium*, which would become the model for numerous similar works in Latin and vernacular languages, the section on any given plant is divided into sections labelled 'Names' (*nomina*), 'Types or species' (*genera*), 'Shapes' (*forma*), 'Place' (*locus*), and 'Time' (*tempus*).[57] 'Time' is clearly one of the main factors of plant being according to early modern botany. Looking at the original Latin as well as the French translation of Fuchs, which appeared before Ronsard's ode, we find copious information about the *time* of different kinds of plants. Thus, 'absinthe [...] must be picked in the month of July'.[58] We must 'gather' (in French *cueillir*—the word that Ronsard uses) the marshmallow roots 'towards the end of August or in the beginning of September', whereas that plant's 'leaves and seeds must be collected only in the summer'. The plant flowers, finally, 'in the months of July and August'. *To be a marshmallow plant is to follow this temporality.* To be a chamomile plant is to follow a different *tempus*: it 'can—in warm climes—be picked [*se cueille*] in springtime; but

54 Theophrastus, *Enquiry Into Plants*, IV. 13, 1–2.

55 Ibid.

56 Ibid., IV, 13, 3.

57 This is the structure used in all entries of Fuchs, *De Historia stirpium*.

58 Fuchs, *Commentaires tres excellens de l'hystoire des plantes*, fols Lvi r–M r and Fuchs, *De Historia stirpium*, pp. 656–658.

in these [our] cold regions, picking only occurs at the start of summer, which is to say in the month of June'. As for roses, to bring us wholly within the Ronsardian, we read: 'Just as the rose appears the latest amongst the beautiful flowers of spring, so it is the first to pass. *It must thus be picked [cueillir] in the month of June, as soon as it is seen*'[59]. In Fuchs's *Historia stirpium*, the rose is the plant-that-requires-urgent-plucking. Not only does Fuchs state at which moment of the year (i.e. in June) we should pick roses, he also underscores how quickly we must pick it: 'Ne mox nusquam c[om] pareat!' 'Dès incontinent qu'on l'apperçoit!' ('Pick it as soon as you see it!') Behind the so-called *carpe diem* motif in Ronsard's ode there is thus, in essence, a *carpe florem* sense of plant time.

On the historicist front, one might thus posit that Ronsard's ode deploys the *carpe diem* motif in light of the botanical reality of the quick passing of roses, of which readers of Fuchs and other writers—not to mention gardeners—would likely have been aware. It is a fair assumption that, given the evolution of sixteenth-century botany, had there been a poem identical to Ronsard's two hundred years earlier, its words might not have resonated in the same way. On the *desencontro* front, the poem's botanical correspondence with regard to time reminds us to see a *plant* here, to meet it, somehow—and it also leaves us wondering about the directionality of the human/plant comparison. As humans, we know a lot about human mortality, and feel it intimately. Do we need a rose to teach us that? Perhaps we need a rose to *feel* that—and this feeling is perhaps, as Irigaray and Marder would put it, communion. What if we read the poem the other way round? What if our own firsthand, existential, anguished awareness of human mortality is what allows us to see the passing of the rose? Indeed, to see the rose at all? We might then see the rose in the poem not (only) as a symbol of *our* aging and death, but as a rose with which we commune because we *already know* that life passes quickly.

Conclusion

In this chapter, by mediating between an ode, its reception history, early modern botany, and plant theory, the goal has been to examine how, when we see certain signifiers for plants (e.g. the word *rose*), it is all too easy to *not actually appreciate that plant's plantness*. We can *read* the word 'rose' and—because of its simplicity, its familiarity, its poetic-ness, its inclusion in

59 Fuchs, *Commentaires tres excellens de l'hystoire des plantes*, fol CCLIIIIr.

a thousand tropes and similes—skip on past it towards that of which it is a symbol, an allegory, an illustration. The preceding pages do not offer a new reading of Ronsard's ode, whose aim it would be to stop us from appreciating the *carpe diem* motif. They call, rather, for an open reading, one whose *debole* grasp is comparable to that of the claw crane merchandiser games we find in arcades, whose claws most often touch but fail to pick up the coveted cuddly toy. They call for allowing the poem to be a *desencontro*, in which the botanical *carpe rosam* and the human *carpe diem* give meaning to one another. Such a reading is an un-reading, perhaps; as Timothy Morton has put it, a 'functional definition of an adult book is one in which nonhumans don't speak and aren't on an equal footing with humans', such that young adult and adult Literature with a capital-L is often fodder for the 'anthropocentrist in training'.[60] A *good* education means that by the age of ten, children 'have already decided that literature should not be about talking toasters of friendly frogs'.[61] In other words, learning to read often means splitting off the material reality of humans and nonhumans living together from the correlationist *for-me* world, in which toasters, frogs, and roses, when they circulate in signs, are there only to furnish elements for understanding human life. The un-reading in the preceding pages looked to both historicism (the reception history of the ode; the potential echoes of Ronsard's poem in Charles de L'Écluse's French version of Dodoens and Guillaume Guéroult's translation of Fuchs) *and* to theory (especially that of Marder and Irigaray) to bring into focus, side by side, both the cultural grasp on Ronsard's ode *and* the plant-ness of the rose. The ode, a product of Culture, here reveals itself—if we let it—as offering us time with what was once called Nature (i.e. the physical world), and more specifically with the vegetal. It offers us a moment of communion with the living, a brief escape from human exceptionalism, something more (and less) than a famous poem by the prince of poets.

Works Cited

Ausonius, *Works*, ed. and trans. by Hugh G. Evelyn-White (Cambridge, MA: Harvard University Press, coll. Loeb, 1921).

Barbier, Jean-Paul, *Ma Bibliothèque poétique* (2: *Ronsard*) (Geneva: Droz, 2017).

Bedini, Gianni, *L'Orto Botanico di Pisa: Piante, storia, personaggi, ruoli* (Pisa: Plus, 2007).

60 Morton, *Humankind*, p. 14.

61 Ibid., p. 15.

Brancher, Dominique, *Quand l'esprit vient aux plantes: botanique sensible et subversion libertine (XVIe-XVIIe siècles)* (Geneva: Droz, 2015).

Chardavoine, Jehan, *Le recueil des plus belles et excellentes chansons en forme de voix de ville, tirees de divers autheurs & poëtes françois, tant anciens que modernes. Ausquelles a esté nouvellement adapté la musique de leur chant commun, à fin que chacun les puisse chanter en tout endroit qu'il se trouvera, tant de voix que sur les instruments* (Paris: chez Claude Micard, au clos Bruneau, à l'enseigne de la Chaire, 1576).

Coccia, Emanuele, *La Vie des plantes. Une métaphysique du mélange* (Paris: Payot et Rivages, 2016).

Collarile, Luigi, *Ronsard et la mise en musique des* Amours *(1552–1553)* (Paris: Classiques Garnier, 2016).

Cuadraro, J. Martinez, 'Ronsard en el arco tensado entre Ausonio y Guillen' in *Actos del V coloquio de la Asociación de Profesores de filología francesa de las universadades* (Murcia, 1996), pp. 315–324.

Descola, Philippe, *Par-delà nature et culture* (Paris: Gallimard, 2005).

Dodoens, Rembert, *Histoire des plantes, en laquelle est contenue la description entière des herbes* [...] *non seulement de celles qui croissent en ce païs, mais aussi des autres estrangères qui viennent en usage de médecine* trans. by Charles de l'Écluse. (Antwerp: impr. de J. Loe, 1557).

Creore, A.E., *A Word-Index to the Poetic Works of Ronsard* (Leeds: W.S. Maney and Son Ltd., 1972).

Dulmet, Florica, 'Ronsard, poète des roses, des femmes et des princes' in *Écrits de Paris. Revue des questions actuelles* (September 1985), 64–71.

Duport, Danièle, *Les Jardins qui sentent le sauvage. Ronsard et la poétique du paysage* (Geneva: Droz, 2000).

Egmond, Florike, *The World of Carolus Clusius: Natural History in the Making, 1550–1610* (London: Pickering & Chatto, 2010).

———, *Eye for Detail: Images of Plants and Animals in Art and Science, 1500–1630* (London: Reaktion Books, 2017).

Egmond, Florike, Paul Hoftijer, and Robert Visser, eds., *Carolus Clusius: Toward a Cultural History of a Renaissance Naturalist* (Amsterdam: Koninklije Nederlandse Akademie van Wetenschappen, 2007).

Findlen, Paula, *Possessing Nature: Museums, Collecting, and Scientific Culture in Early Modern Italy* (Berkeley: University of California Press, 1994).

Fuchs, Leonhart, *De historia stirpium commentarii insignes* [...] *Una cum quadruplici Indice* [...] *maximis impensis et vigiliis elaborati, adiectis earundem vivis plusquam quingentis imaginibus, nunquam antea ad naturae imitationem artificiosius effictis & expressis, Leonharto Fuchsio medico hac nostra aetate longe clarissimo, autore.* (Basileae: in officina Isingriniana, anno Christi 1542).

———, *Commentaires tres excellens de l'hystoire des plantes composez premierement en latin par Leonarth Fousch, medecin tres renommé: Et depuis, nouvellement traduictz en langue françoise, par un homme scavant & bien expert en la matiere* (Paris: chez Jacques Gazeau, en la rue Sainct Jehan de Latran, devant le college de Cambray, 1547–1548). [BnF Arsenal FOL-S-560]

Greene, Edward Lee, *Landmarks of Botanical History*, ed. by Frank N. Egerton (Stanford: Stanford University Press, 1983).

Hall, Matthew, *Plants as People: A Philosophical Botany* (New York: SUNY Press, 2011).

Halévy, Olivier, and Jean Vignes, 'Paris, 1553 : Un exemple de "microhistoire littéraire"', *Revue de la BNF* 33.3 (2009), 64–71.

Haraway, Donna J., *Companion Species Manifesto: Dogs, People, and Significant Otherness* (Chicago: Prickly Paradigm Press, 2003).

Irigaray, Luce, *J'aime à toi: Esquisse d'une félicité dans l'histoire* (Paris: Grasset, 1992).

———and Michael Marder, *Through Vegetal Being: Two Philosophical Perspectives* (New York: Columbia University Press, 2016).

Johnson, Leonard W., 'La salade tourangelle de Ronsard' in *Littérature et gastronomie*, ed. by Ronald W. Tobin (Tübingen: Biblio 17, 1985), pp. 151–171.

Kusukawa, Sachiko, *Picturing the Book of Nature: Image, Text, and Argument in Sixteenth-Century Human Anatomy and Medical Botany* (Chicago: University of Chicago Press, 2012).

Labaste, Jacqueline, 'Un nouveau livre de la bibliothèque de Ronsard', *Bibliothèque d'Humanisme et Renaissance* (*BHR*) 47 (1985), 613–618.

La Cresme des bons vers, triez du meslange & cabinet des sieurs de Ronsard, dv Perron, de Malerbe, de Sigongnes, de Lingendes, Motin, Maynard, de Bellan, d'Vrfé, Theophile & autres (Lyon: M. Covrant, 1622). [Columbia University Rare Book and Manuscript Library, call number: B843.11 C86]

Lafont, Suzanne, 'Rose, femme, événement: parcours d'un poncif', *Poétique* 106 (April 1996), 179–197.

Latour, Bruno, *Nous n'avons jamais été modernes* (Paris: La Découverte, 1991).

Laumonnier, Paul, *Ronsard, poète lyrique* (Paris: Hachette, 1909).

———, 'Sur la bibliothèque de Ronsard', *Revue du XVIe siècle* (1927), 315–335.

Livet, Henri-Philippe, 'Sur le rosier: Ronsard fleurit la France', *Arts et idées. Revue bimestrielle* 2.12 (December 1937), 21–25.

Livingston, Charles H., 'Notes sur la bibliothèque de Ronsard', *Bibliothèque d'Humanisme et Renaissance* (*BHR*) 23 (1961), 53–57.

Marder, Michael, *Plant-Thinking: A Philosophy of Vegetal Life* (New York: Columbia University Press, 2013).

———, *Philosopher's Plant: An Intellectual Herbarium* (New York: Columbia University Press, 2014).

Morton, Timothy, *Humankind: Solidarity with Nonhuman People* (New York: Verso, 2017).

Nealon, Jeffrey T., *Plant Theory: Biopower and Vegetable Life* (Stanford: Stanford University Press, 2016).

Ogilvie, Brian W., *The Science of Describing: Natural History in Renaissance Europe* (Chicago: University of Chicago Press, 2008).

Ondra, Nancy J., *Taylor's Guide to Roses: How to Select, Grow, and Enjoy More Than 380 Roses* (New York: Houghton Mifflin, 2001).

Rioux, Jean-Antoine, ed., *Le Jardin des plantes de Montpellier: quatre siècles d'histoire* (Montpellier: Sauramps médical, 2014).

Rossi, Michel, *Le Jardin des plantes de Montpellier: de la médecine à la botanique* (Cersailles: Éditions Quae, 2013).

Ronsard, Pierre de, *Les Amours de P. de Ronsard Vandomois, nouvellement augmentées par lui, & commentées par Marc Antoine de Muret. Plus quelques Odes de L'auteur, non encor imprimées* (Paris: Maurice de la Porte [veuve], 1553).

———, *Les Œuvres de Pierre de Ronsard* (Paris: Nicolas Buon, 1623).

———, *Œuvres complètes*, ed. Paul Laumonier (Paris: Hachette, 1914–1967).

———, *Selected Poems*, ed. and trans. by Malcolm Quainton and Elizabeth Vinestock (London: Penguin, 2002).

Rouget, François, *Ronsard et le livre: étude de critique génétique et d'histoire littéraire* (Geneva: Droz, 2010–2012).

———, ed., *Dictionnaire de Pierre de Ronsard* (Paris: Champion, 2015).

Selosse, Philippe, 'Traduire les termes de couleurs: la chromonymie en botanique à la Renaissance', in *Mots, termes et contextes*, ed. by Daniel Blampain, Philippe Thoiron and Marc Van Campenhoudt (Paris: Éditions des Archives Contemporaines, 2006), pp. 619–629

Shapiro, Norman R., *Lyrics of the French Renaissance: Marot, Du Bellay and Ronsard* (New Haven: Yale University Press, 2002).

Simonin, Michel, '"Poésie est un pré", "Poème est une fleur": métaphore horticole et imaginaire du texte à la Renaissance', in *L'Encre et la lumière: quarante-sept articles (1976–2000)*, ed. by Michel Simonin (Geneva: Droz, 2004), pp. 253–263.

Theophrastus, *Enquiry Into Plants*, trans. by Arthur Hort (Cambridge: Harvard University Press, coll. Loeb Library, 1916).

Thibault, Geneviève, and Louis Perceau, *Bibliographie des poésies de P. de Ronsard mises en musique au XVIe siècle* (Paris: Droz, 1941).

Usher, Phillip John, *Exterranean: Extraction in the Humanist Anthropocene* (New York: Fordham University Press, 2019).

Vattimo, Gianni, and Pier Aldo Rovatti, *Il Pensiero debole* (Milan: Feltrinelli, 1988).

Veyrin-Forrer, Jeanne, 'La bibliothèque de Ronsard', in *Ronsard, la trompette et la lyre* (Paris: Bibliothèque nationale de France, 1985), pp. 165–169.

Yandell, Cathy, *Carpe Corpus: Time and Gender in Early Modern France.* (Newark: University of Delaware Press, 2000).

About the Author

Phillip John Usher Associate Professor of French Literature, Thought and Culture and of Comparative Literature at New York University, where he is also Chair of the Department of French Literature, Thought and Culture. He is the author, translator, or editor of nine volumes. His most recent monograph is *Exterranean: Extraction in the Humanist Anthropocene* (New York: Fordham University Press, 2019).

8. Renascent Nature in the Ruins: Joachim du Bellay's *Antiquitez de Rome*

Victor Velázquez

Abstract

Joachim du Bellay's *Les Antiquitez de Rome* (1558) is traditionally read as a text about human-made culture: the grandeur and ruin of Rome. Nevertheless, through a moral condemnation of imperial Rome's pride and its violent origins, Du Bellay describes the effects Rome's fall had on the nonhuman landscape, thus inviting a re-evaluation of the relation between humans and nonhuman nature. His juxtaposition of the destructiveness of history's blindness to nature with the landscape's re-emergence from the ruined remains of Roman culture yields images that challenge us to rethink conservation in relation to a nature that changes over time, and which is inseparable from culture and its ruins, while at the same time redefining the traditional presupposition of what we categorize as 'nature writing'.

Keywords: Du Bellay, ruins, pride, time, natural cycle, regeneration

The intersection of technology, power, and hubris in the modern era has created a context in which we must rethink the relationship between human and nonhuman natures and cultures. Both the Anthropocene (according to one of its possible start dates) and the ecological turn in the study of literature can be situated in relation to the first detonation of nuclear weapons.[1] According to some, our current epoch thus began in 1945, while more generally Gabriel Egan argues that environmentalism is 'a response to the rapid increase in the power of human technologies and the hubris of the scientists and technocrats in charge of them', with the first atomic test carried out by Oppenheimer

1 Lewis and Maslin, 'Defining the Anthropocene'; Rickards, 'Metaphor and the Anthropocene', p. 281.

Goul, P. and P.J. Usher (eds.), *Early Modern Écologies. Beyond English Ecocriticism.*
Taylor & Francis Group, 2020
DOI 10.5117/9789462985971_CH08

and Segrè being selected as a turning point: 'it carried a small, but quite real, chance (about one-in-fifty, some of them thought) of instantly igniting the world, and they decided to risk it'.[2] Turning to sixteenth-century French texts in this context can help us reconceptualise the relation between humanity, human culture, and nonhuman nature. Some might resist such a move: in *Forests: Shadow of Civilization*, Robert Harrison argues that humanism divorces the human from the nonhuman, such that humans dominate nature.[3] But claims such as that one merely confuse humanism with anthropocentrism.[4] Early modern texts frequently explore the limits of humanity as well as humanity's place within the cosmos. As Erica Fudge shows in *Renaissance Beasts*, for example, humanists' explorations of what appear to be purely anthropocentric issues such as politics and law were often intricately related to inquiries into nonhuman nature.[5] In what follows, I shall argue that a new reading of Joachim du Bellay's meditation on the paradox of Rome's grandeur and ruin in his collection of sonnets, *Les Antiquitez de Rome* (1558), can make two important contributions to the ecocritical conversation: 1) it provides a description of the demise of a highly militarized and arrogant civilization and of the unexpected effects its fall had on the nonhuman landscape; 2) it invites an analysis of the challenge of conservation in relation to the destructiveness of time through a meditation on the ruins of human-made artefacts and the culture in which they were created, but which nonetheless lends itself to a reflection on the conservation of nonhuman nature. Du Bellay's text might thus help us reframe how we conceptualize the questions of conservation and the relation between human and nonhuman nature within the context of an ecocritical discourse. At the same time, the *Antiquitez* challenge the traditional presupposition of what might be considered 'nature writing': Du Bellay's sonnet sequence has traditionally been read almost exclusively as a text about human-made culture, particularly the 'Eternal city', the Roman Empire and the poetic enterprise. And yet, this sequence of poems offers a surprising meditation on Nature, which is articulated through a moral argument that describes the effects of Roman culture and pride on the Italian landscape. In a sense, the lyric sequence asks us to follow Timothy Morton's suggestion to slow down and question the deeper relation between apparently 'technocultural-aesthetic' issues and 'wet and organic' ones.[6]

2 Egan, *Green Shakespeare*, p. 18; p. 17.

3 Harrison, *Forests*, p. 92.

4 Gouwens, 'What Posthumanism Isn't'; Usher, 'Untranslating the Anthropocene'.

5 Fudge, *Renaissance Beasts*, p. 3.

6 See *Ecology Without Nature*, p. 28.

'*De la mort vient la renaissance*' ('From death comes rebirth')[7]

A first wave of ecocriticism placed emphasis on the representation of so-called natural places (rivers, watersheds, forests, etc.) in nature writing. And yet, thinking of our life in nature ought to cause us to consider the very ground of our being not only in 'natural spaces', but also within culture (i.e. constructed places) and time. Although some critics have signalled that nature is simultaneously both in place and in time, the exploration of temporality as a central aspect of ecocriticism has been largely overlooked.[8] In contrast, Renaissance writers, particularly Humanists, would have been familiar with the Ancient Greek notion *physis*, that is, a nature that is conditioned by time: the process of birth, growth, and decay.[9] Moreover, as Andrew Hui points out, it is precisely in the Renaissance that writers became acutely aware of the passage of time in a human and historical sense as well: the material remains of antiquity that were being recovered testified both to the greatness of past ages as well as to how much of the past was lost, and the difficulty of recovering it: 'Either the achievements of the ancients are so great that it is impossible to equal them, or the achievements of the ancients are so fragmented that it is impossible to know them'.[10] Accordingly, the concept of the passage of time and the anxiety over its destructiveness, represented through the *carpe diem* and *exegi monumentum* motifs, seem to be an obsession in many Renaissance texts, and especially so in lyric poetry.[11] Du Bellay's sonnets are no exception. As George Hugo Tucker notes, Du Bellay's meditation on Rome's ruins situates itself in a long tradition that attempts to interpret the interrelation of the destructiveness of time and the potential survival of the Eternal City even in its ruin.[12]

7 Belleau, 'La Pierre Lunaire', p. 141.

8 See the Introduction to Merrill Ingram's *Coming into Contact*.

9 Although Plato's understanding of the natural world as a representation of a 'true nature' that exists in the metaphysical sphere has had an important resonance for Western culture's understanding of the physical world—especially as his perspective was appropriated through the Christian lens—according to Aristotle, *physis* represents the internal principle of change of a living thing that explains its capacity to change over time while retaining its identity (*Phys.* 193b21–22); similarly, as it will be shown below, other Hellenic thinkers such as Heraclitus provided Renaissance writers with an understanding of nature in which temporality was central.

10 Hui, *The Poetics of Ruins in Renaissance Literature*, p. 165.

11 In *Carpe Corpus* Cathy Yandell argues that the temporal topoi in lyric poetry reveal the poets' struggle with the menacing, and ravaging powers of time as well as the desire to control time and even counteract death, p. 23. Indeed, the lyric poem stands out as the poet's yearning for (literary) immortality. See Hui, *The Poetics of Ruins*, p. 18.

12 See Tucker, *The Poet's Odyssey*, especially Chapter 2.

Du Bellay's representation of Rome's ruined architecture at the centre of the sonnet sequence explores the site of the changing of the Roman landscape over time. Indeed, the site itself (from the Latin *situ*, which means both site and decay)[13] betrays the effects of time as well as a complex relation between nonhuman nature and human-made cultural artefacts. In the eighteenth sonnet, for example, the reader is presented with the transition from a pastoral landscape to a monumental city marked by great palaces that, as the first verse highlights, are eventually reduced to a heap of rocks:

Ces grands monceaux pierreux, ces vieux murs que tu vois,
Furent premierement le cloz d'un lieu champestre:
Et ces braves palais dont le temps s'est fait maistre,
Cassines de pasteurs ont esté quelquefois. (1–4).

(These heapes of stones, these old wals which ye see,
Were first enclosures but of saluage soyle;
And these braue Pallaces which maystred bee,
Of time, were shepheards Kingly ornaments).[14]

This initial quatrain points the reader towards the natural process of birth, growth, and death of the Roman Empire. However, it does so in a peculiar order: the sonnet starts with an image of the death of the Empire, that is, the 'present' state of ruins and rubble at the time of Du Bellay's visit to Rome. Although largely absent in their regal wholeness from the sonnet, the construction of those 'brave palais' ('elegant palaces') took place at the expense of pastoral landscapes. Only fragments of the grandeur of Rome remain. In *The Vision of Rome in Late Renaissance France*, McGowan shows that for Du Bellay (especially in the *Deffence*) the fragments of Rome are seen as inadequate in and of themselves for reconstructing a coherent whole.[15] Accordingly, these buildings are always presented in the sonnet as ruined and lost to the past. They signal their absence much more than their former presence. Thus they suggest a return to the un-built, towards a certain definition of nature; or rather, they challenge us to rethink the relation of human-made artefacts and the nonhuman nature in which they appear. The palaces lose their shape and meaning, and litter the natural landscape

13 See Hui, *The Poetics of Ruins*, pp. 35–36.

14 Unless otherwise noted, I will use to Edmund Spencer's English translation of the *Antiquitez*. 'Ruines of Rome' in *The Works of Edmund Spenser: The Minor Poems, Volume Two*.

15 McGowan, *The Vision of Rome in Renaissance France*, p. 193.

in the form of heaps of stones at the same time that they become part of that same nonhuman, 'natural' landscape. And yet, in the poet's vision, this formless rubble resembles the shape of enclosures of an uncultivated, proto-pastoral field. Such a vision hints at the impossibility of thinking of a pristine nature before culture.[16]

Ironically, the decaying ruins of the empire also inspire a birth narrative that is grounded in the natural landscape.[17] As McGowan notes, the fragmentary nature of the Roman ruins allowed the artists of that time to enter into a project of reconstruction of the whole.[18] For Du Bellay, the presence of the 'heapes of stones', seems to evoke the mythic origin narrative of Rome as it focuses on the landscape's change over time. The quatrain recalls, through its references to the fields and pastures of times past, the prehistory of the Roman Empire by drawing to mind King Evander's primitive pastoral settlement as described in Book 8 of the *Aeneid*. Such a reversal in the narrative—starting from death and decay and continuing with new life—not only points toward the possible renewal of an empire,[19] but also seems to emphasize the rebirth of nature within the ruins of Rome. McGowan suggests as much when she describes the ambiguous vision of Rome before Du Bellay: 'Giant structures are mutilated; theatres are sad and silent; walls represent an undifferentiated mass *as herbs and wild grasses cover their surface and screen their meaning*'.[20] In a similar way, the impossibility of conserving the cultural artefacts of Rome also creates an opening for a re-emergence of human culture through the art of Du Bellay's lyric poetry. In order for Du Bellay's sonnets to emerge from the smoldering ashes of antiquity, 'he willed, dreamed, and needed Rome to be in ruins'.[21]

16 Following Morton's view, in *The Literature of Waste* Susan Morrison suggests that such a concept of nature 'perpetuates the division between human and nature' and reinforces anthropocentricism, p. 123.

17 McGowan notes that '[t]he value [of fragments] lies in the way broken or unfinished fragments expose the technical process which went into their making' (*The Vision of Rome*, p. 194). Here we note how the fragments literally evoke the process of both decay and growth. Similarly, Hui argues that if the Roman ruins evoke the dissolution of meaning and incoherence, Renaissance poetics strived to give coherence that would give shape to them and yield a new poetic monument, p. 163.

18 McGowan, *The Vision of Rome*, pp. 163ff.

19 As many critics note, one of the main goals of the *Antiquitez* is to establish France as the new Rome both politically and within the literary tradition. This aspect of the sonnets is most explicit in *Au Roy*: 'Que vous puissent les Dieux un jour donner tant d'heur, | De rebastir en France une telle grandeur | Que je la voudrois bien peindre en vostre langage', pp. 9–11. ('May the Gods one day give you the good fortune, | To rebuild in France such greatness | That I would willingly paint it in your language'; translation by Helgerson).

20 McGowan, *The Vision of Rome in Renaissance France*, p. 189, emphasis added.

21 Hui, *The Poetics of Ruins*, p. 145.

The sonnets at the heart of the collection capture and amplify this intermingling of nature and culture by depicting ruins colonized by a renewed nature, and foregrounding the presence of the wilderness springing out of the rubble.[22] Or rather, the ruins join nonhuman nature inasmuch as the difference between themselves and the undifferentiated wilderness in which they are situated is blurred. As the shapeless mass represented by the ruins of Roman culture[23] becomes absorbed into the natural landscape, it suggests that the building and decomposition of human structures does not happen outside of nature:[24] 'Nature has transformed the work of art into material for her own expression as she had previously served as material for art.'[25]

Starting with the fourteenth sonnet, an extended simile is inhabited with images referring not only to temporal markers and pastoral landscapes, but also to wild beasts:

Comme on passe en æsté le torrent sans danger,
Qui souloit en hyver estre roy de la plaine,
Et ravir par les champs d'une fuite hautaine
L'espoir du laboureur, & l'espoir du berger.
Comme on void les coüards animaux oultrager
Le courageux lyon gisant dessus l'arene,
Ensanglanter leurs dents, & d'une audace vaine
Provoquer l'ennemy qui ne se peult vanger. (1–8).

(As men in Summer fearless passe the foord,
Which is in Winter lord of all the plaine,
And with his tumbling streames doth beare aboord,
The ploughmans hope, and shepheards labour vaine:
And as the coward beasts vse to despise,
The noble Lion after his liues end,
Whetting their teeth, and with vaine foolhardise
Daring the foe that cannot him defend).

22 For a nuanced study of the difficulties in preserving cultural artifacts from the ravages of time, see Desilvey, *Curated Decay: Heritage Beyond Saving.*

23 Hui, *The Poetics of Ruins*, p. 159.

24 Accordingly, Hui suggests that ruins are a 'natural part of the life cycle of a building'. *Poetics of Ruins*, p. 75.

25 Gerog Simmel, cited in Hui, *Peotics of Ruins*, p. 60. Similarly, Morton suggests, culture does not happen outside of nature, or even at its expense, but they are rather intricately interwoven without necessarily collapsing into an identity. See *Ecology Without Nature*, p. 21.

Rome's royal greatness and might is identified as a tremendously devastating force by the verb *ravir*. But here the destructive power is evoked as having passed by the use of the imperfect tense ('souloit'). At the time of enunciation, the greatness of old is no-longer present, and neither is its destructive force. The first verse places emphasis on that by noting the lack of danger using the present tense 'on passe [...] sans danger'. It is worth noting that the destruction of the empire is focused on the landscape: it is the 'plaine', the 'champs', and the harvest of the field worker that are subject to destruction.

The opening images of the fourteenth sonnet recontextualize the history of Rome in terms of seasons and the natural phenomena that coincide with them. The experience of time as a cyclical movement that enveloped human activity within natural rhythms was a novel idea in the Renaissance, and it was during this period that 'the course of human life became identified with vegetable growth [...] [and] historical developments were also viewed in the same light'.[26] Within the context of temporality and cycles, the transition from pastures in the first quatrain to the image of predatory animals in the second, seems to suggest a return to a chaotic new kind of wilderness—one in which the meek mock the proud. The tercets of the sonnet also incorporate this cyclical theme that elaborates the reversal of strength and weakness. Du Bellay presents the pre-history of Rome with his description of Achilles and the Greek camp's abuse of Hector's dead body:

> Et comme devant Troye on vid des Grecz encore
> Braver les moins vaillans autour du corps d'Hector:
> Ainsi ceulx qui jadis souloient, à teste basse,
> Du triomphe Romain la gloire accompagner,
> Sur ces pouldreux tumbeaux excercent leur audace,
> Et osent les vaincuz les vainqueurs desdaigner. (9–14).

> (And as *Troy* most dastards of the Greekes
> Did braue about the corpes of *Hector* colde;
> So those which whilome wont with pallid cheeks
> The Romane triumphs glorie to behold,
> Now on these ashie tombes shew boldnesse vaine,
> And conquer'd dare the Conquerour disdaine).

The verses present a repetitive narrative that seems to point simultaneously to the beginning and to the end of the Roman Empire. The inversion of power

26 Glasser, *Time in French Life and Thought*, p. 179.

that is evident as the formerly conquered people show disdain for their Roman conqueror evokes the fall of Troy; and yet, as the image of Troy's fall is called to mind, it cannot but recall the establishment of Rome. Echoing the recontextualization of the history of Rome within a seasonal context, the reader is alerted to the imminent return of a great power: even if just now we are in '[l']æsté [...] sans danger', we are also reminded that the ravages of '[l']hyver' will return; if Troy's military strength is vanquished with the fall of Hector, we are also reminded that the reign of Roman Caesars was also to come.[27] Du Bellay's use of a naturalizing trope for explaining Rome's rise and fall identifies the natural cycles of time and the life cycle itself as inseparable from the anthropocentric actions of humanity in history. In addition, within that context, the human endeavours seem to become subjugated to the powers of nature and time.[28]

A just end: *le 'uieil peché'* (the 'ancient sin') and its natural consequences

The eighteenth and sixteenth sonnets, which as Tucker points out are placed diagonally opposite one another at the centre of the collection in the original 1558 edition, likewise locate a triumphal, renascent nature at the centre of Rome's ruins. The extended comparison of the sixteenth sonnet depicts the cycle of Roman power in the almost sublime terms of a return to an original, pre-human, natural chaotic state. It is worth reproducing the sonnet in its entirety:

> Comme lon void de loign sur la mer courroucee
> Une montagne d'eau d'un grand branle ondoyant,
> Puis trainant mille flotz, d'un gros choc abboyant

27 Moreover, the reader is also alerted to the potential future of re-emergence of the empire under the rule of France. See note five above.

28 Gadoffre sees in this a counterpoint movement: 'A travers les affabulations romaines on voit poindre une protestation solennelle contre la corruption et "perte de la mémoire du Temps", une affirmation de l'existence d'un cosmos créé et périssable, et d'une histoire humaine faite de cycles d'ascensions et de décadences, d'un immense contrepoint qui fait jouer l'un contre l'autre le temps astronomique, le temps historique et les cycles de la végétation' (*Du Bellay et le sacré*, p. 144). Through the Roman myths one sees a solemn declaration against corruption and the 'loss of the memory of Time' emerge, an affirmation of the existence of a created and perishable cosmos, and of a human history made of cycles of upward and downward movements, of an immense counterpoint that puts astronomic time, historic time and the cycles of vegetation in play with one another.

Se crever contre un roc, où le vent l'a poussee,
Comme on voit la fureur par l'Aquilon chasee
D'un sifflement aigu l'orage tournoyant,
Puis d'un ælle plus large en l'air s'esbanoyant
Arrester tout à coup sa carriere lasee :
Et comme on void la flamme ondoyant en cent lieux
Se rassemblant en un, s'aguiser vers les cieux,
Puis tumber languissante: ainsi parmy le monde
Erra la monarchie: & croissant tout ainsi
Qu'un flot, qu'un vent, qu'un feu, sa course vagabonde
Par un arrest fatal s'est venue perdre icy.

(Like as ye see the wrathfull Sea from farre,
In a great mountaine heap't with hideous noyse,
Eftsoones of thousand billowes shouldred narre,
Against a Rocke to breake with dreadfull poyse:
Like as ye see fell *Boreas* with sharpe blast,
Tossing huge tempests through the troubled skie,
Eftsoones hauing his wide wings spent in wast,
To stop his wearie cariere suddenly:
And as ye see huge flames spread diuerslie,
Gathered in one vp to the heauens to spyre,
Eftsoones consum'd to fall downe feebily:
So whilom did this Monarchie aspyre
As waues, as winde, as fire spred ouer all,
Till it by fatal doome adowne did fall).

Here, just as in the fourteenth sonnet, the simile suggests a direct relation between nature and culture in the cyclical aspect of their rise and fall. However, this dynamic, cyclical relation of life and death, which appears to be inseparable from nature and culture, is not without a moral interpretation.[29] The eventual, 'arrest fatal' is ultimately precipitated by the arrogant erring of the monarchy ('erra' from the Latin *erro*: to wander, or go astray, or to miss, from which we derive the word 'error') which led it to aspire to reach heaven ('s'aguiser vers les cieux'). The poet seems to identify a divine logic to this kind of life cycle in the eighteenth sonnet:

29 For Morton, it is this realization of our complicity and guilt that dispels the illusion of environmentalism: 'The beautiful soul is dissolved when we recognize that we did it, we caused the environmental destruction, not *you*, whoever you are' (*Ecology Without Nature*, p. 185).

Mais le Ciel, s'opposant à tel accroissement,
Mist ce pouvoir es mains du successeur de Pierre,
Qui sous nom de pasteur, fatal à ceste terre,
Monstre que tout retourne à son commencement. (11–14).

(Till th'heuen it selfe opposing gainst her might,
Her power to *Peters* successor betooke;
Who shepheardlike, (as fates the same foreseeing)
Doth shew, that all things turne to their first being).

The return to the 'commencement' is linked to the wilful intervention of divine forces (le Ciel) that oppose the rise (accroissement) of the Roman Empire. Similarly, in the sixteenth sonnet cited above, the 'arrest fatal' seems to indicate a divine intervention that brings down the haughty: the growth in power of both empire and elements signalled by the descriptive 'croissant' is counteracted with a downward movement to the ground ('à ceste terre').[30]

The poem furthermore indicates that the growth of power in Rome was accompanied by corruption. The characters that could be most closely identified with nature, the 'bergers' and field 'laboureurs', grow in power as they become more civilized by putting on royal garments and arms respectively. A consequence of such a transition from pastoral life to empire building is that freedom was gradually replaced by despotism:

Puis l'annuel pouvoir le plus grand se vid estre,
Et fut encor plus grand le pouvoir de six mois:
Qui, fait perpetuel, creut en telle puissance,
Que l'aigle Imperial de luy print naissance. (7–10).

(Eftsoones their rule of yearely Presidents
Grew great, and sixe months greater a great deele;
Which made perpetuall, rose to so great might,
That thence th'Imperiall Eagle rooting tooke).

30 Tucker argues that Du Bellay attributes the cause of Rome's downfall not only to the excessive human ambition but also to divine jealousy (*The Poet's Odyssey*, p. 57). Likewise, Gadoffre suggests that in Du Bellay's *Antiquitez,* the fall of Rome is best understood as divine punishment for an original sin (*Du Bellay et le sacré*, pp. 128–129).

Thus the poem not only situates nature and culture within the context of temporality and life cycles, but also in relation to ethics and morality.[31] The image of Roman ruins would doubtlessly recall the eschatological themes of death, judgment and condemnation to its primarily Christian readers. Although Tucker suggests that there is nothing that particularly encourages the reader to interpret the *Antiquitez*'s treatment of morality, specifically the 'uieil peché' cited in the twenty-fourth sonnet, in the light of its Biblical resonances,[32] it is significant to note that, at least historically, that has not been the way the text was received. Dutch humanist Jan van der Noodt, for example, interpreted the last fifteen sonnets of the collection precisely in light of their biblical resonances when he used them as apocalyptic emblems.[33] Gadoffre similarly contends that Du Bellay's work can best be understood as 'un code théologique superposé au code humaniste' ('a theological code superimposed over a humanist one').[34] Whether intentional or not, Du Bellay's description of Rome's sin(s) and its consequences lend themselves to such readings. In the third sonnet the sin of pride is suggested as the cause of the city's destruction:

> Voy quel orgueil, quelle ruine: & comme
> Celle qui mist le monde sous ses loix,
> Pour donter tout, se donta quelquefois. (5–7).
>
> (Behold what wreake, what ruine, and what wast,
> And how that she, which with her mightie powre
> Tam'd all the world, hath tam'd herselfe at last).

The moral judgment of Rome's arrogance is highlighted precisely by the emphasis on dominion, since it is the desire to dominate that clearly entails the destruction of the empire.[35] Within the context of the entire sonnet, these

31 As Tucker notes, 'Du Bellay's guide-like presentation [...] of the Roman remains to the "nouueau uenu" is *par excellence* the product of a (twofold) textual meditation (through Vitalis in 1553 and 1554) as opposed to direct personal observation. The resulting "picture", abstract and moral rather than concretely visual or touristic is typical of that larger "tableau" which is *Le Premier livre des Antiquitez de Rome*' (*The Poet's Odyssey*, pp. 110–111).

32 Tucker links Du Bellay's representation of Rome's fall to the ancient Greek and epic traditions, and argues for a 'pagan' reading of the *Antiquitez* in contrast to the *Christian* reading of *Songe*. *The Poet's Odyssey*, pp. 180–181.

33 See Russell, 'Du Bellay's Emblematic Vision of Rome', p. 101.

34 Gadoffre, *Du Bellay et le sacré*, p. 130. English translation is mine.

35 Although the poem's reference is to the history of the Roman Empire that put the known world under its law, we can also read '*Pour donter tout*' as including the dominion over the

lines serve as an epigram reflecting a moral judgment against arrogance as much as an observation of time's effect on physical existence.[36]

Du Bellay's poem, moreover, echoes the moral judgment reserved for Babylon in the book of Revelation. Gadoffre argues that Revelation and its Old Testament intertexts serve as important sources for Du Bellay: 'Le rapprochement avec l'Apocalypse pourrait sembler arbitraire à première vue, et pourtant, dans ces sonnets austères et parfois sibyllins on perçoit ça et là les traces d'une eschatologie, ne fût-ce que dans la manière dont le poète met en scène le Chaos. [...] Sous la plume de Du Bellay l'histoire romaine devient ainsi une parabole du destin de l'Univers et la chute de Rome annonce la fin des temps' ('The comparison with the Apocalypse could seem arbitrary at first sight; however, in these austere and sometimes enigmatic sonnets one perceives here and there the traces of an eschatology, if only in the manner in which the poet stages Chaos [...] Through the pen of Du Bellay Roman history thus becomes a parable of the destiny of the universe and the fall of Rome announces the end of time').[37] The book of Revelation describes Rome as the 'great city that rules over the kings of the earth' (17. 18), just as Du Bellay writes that Rome 'mist le monde sous ses loix' ('placed the world under its laws'). Moreover, the apocalyptic prophet identifies the city's destruction to be the result of its own guilt:

> Give back to her as she has given;
> Pay her back double for what she has done.
> Mix her a double portion from her own cup.
> Give her as much torture and grief
> As the glory and luxury she gave herself. (Revelation 18. 6–7).

In Du Bellay, Rome and its landscape also suffer their own undoing justly, just like the whole Earth—including land, sea and its living creatures—justly suffers the wrath of God in Revelation. As Gadoffre has shown, the concept of an original sin seems to be at the centre of Rome's fated collapse.[38] For example, in the twenty-fourth sonnet, Du Bellay evokes Rome's civil war

natural landscape. As noted above, the eighteenth sonnet's first quatrain suggests an agonistic relation between the erection of the Roman buildings and the landscape. In our contemporary moment, it is worth considering how a national interest of domination might lead to decisions which in the end would lead to devastating effects on the environment.

36 See Melehy, 'Du Bellay's Time in Rome: *The Antiquitez*', p. 9.

37 Gadoffre, *Du Bellay et le sacré*, pp. 140–141.

38 Ibid., p. 129.

and collapses it with the mythical origin of the city through the fratricidal coupling of Romulus/Cain and Remus/Abel:

Vous destrempiez le fer en vos propres entrailles?
Estoit-ce point (Romains) vostre cruel destin,
Ou quelque vieux peché qui d'un discord mutin
Exerçoit contre vous sa vengeance eternelle?
Ne permettant des dieux le juste jugement,
Voz murs ensanglantez par la main fraternelle
Se pouvoir assurer d'un ferme fondement. (8–14).

(Your blades in your owne bowels you embrew'd?
Was this (ye *Romanes*) your hard destinie?
Or some old sinne, whose vnappeased guilt
Powr'd vengeance forth on you eternallie?
Or brothers blood, the which at first was spilt
Vpon your walls, that Godmight not endure,
Vpon the same to set foundations sure?)

Du Bellay identifies human pride and civil war *as well as* wickedness and divine judgment as the reasons for Rome's destruction. The rewriting of Roman history and myth through the Christian concept of sin also allows the image of the bloody fraternal hand to be recontextualized both as an allusion to the immemorial murder of Abel by Cain and as France's own religious turmoil under Henri II. The fall of Rome functions as immemorial history bound to repeat itself in unending displacements, which include Renaissance France and beyond.[39] Du Bellay's Rome enacts such a temporal displacement by telescoping biblical, mythical, and historical referents. His text proposes a moral argument by shadowing the past and casting it as the fated end of the contemporary moment.

It is interesting to consider that Cain is cast as a figure that is extremely at odds with nature and strongly identified with the city: he originally avoids

39 See Gadoffre's assessment of the prismatic vision of history in the *Antiquités*: 'Ce n'est pas seulement de la puissance de Rome qu'il est question ici, mais de la "Monarchie", c'est-à-dire de la notion d'empire universel transmise des Babyloniens aux Mèdes, puis aux Perses, puis aux Macédoniens, puis aux Romains' (p. 49: 'It is not only the power of Rome that is in question here, but that of the "Monarchy", that is to say, the notion of universal empire transmitted from the Babylonians to the Medes, then to the Persians, then to the Macedonians, and then to the Romans'), and then we might add to the list: 'and then to the French', and 'most recently to the Americans'.

working the land and chooses instead to hunt, perhaps already signaling an antagonistic relation to nature; moreover, when he is punished for killing Abel, his curse is that the ground 'will no longer yield its crops for [him]'; finally, he is the first human to build a city in the Bible (Genesis 4. 12–17). Du Bellay's displacement of the history of Rome through the Christian narrative, moreover, seems to run from the beginning to the end of time: form the Fall to the Apocalypse. As such, Rome stands for the sinful human condition[40] and allows the poet to universalize the judgment it incurs: 'Ainsi le monde on peut sur Rome compasser, | Puisque le plan de Rome est la carte du monde' ('Thus one can measure the world by Rome, | Since the plan of Rome is the map of the world') (Sonnet 26: 13–14).

Du Bellay seems to accept the destruction of the Roman Empire as a consequence of its being. When Rome's demise is given meaning at the end of the sequence in the thirty-first sonnet, it is striking that the effect of Rome's judgment is linked with the re-emergence of wilderness: 'De ce qu'on ne void plus qu'une vague campagne | Où tout l'orgueil du monde a veu quelquefois | [...] Tu en es seule cause, ô civile fureur' ('The same is now naught but a champion wide, | Where all this worlds pride once was situate | [...] Thou onely cause, Ciuill furie art') (1–2; 9). And yet there is a tension that remains, which makes the sonnet sequence difficult to interpret in its relation to the question of the permanence of human-made artefacts. As the ending sonnet makes clear, neither architectural nor literary works seem to be able to hope for immortality:

> Si sous le ciel fust quelque eternité,
> Les monuments que je vous ay fait dire,
> Non en papier, mais en marbre & porphyre,
> Eussent gardé leur vive antiquité. (5–8).
>
> (If vnder heauen anie endurance were,
> These moniments, which not in paper writ,

40 The biblical interpretation of original sin proposes that death enters the world as a result of that original disobedience (Genesis 2. 1–617; Romans 5. 12). Through that action all humans assume the guilt of sin and suffer the inevitable consequence: death. Adam and Eve's failure to obey God's commandment not to eat the forbidden fruit of the garden, and not Cain's murder of his brother represents original sin in Judeo-Christian tradition. However, in the sonnet sequence the resonance with Roman legend and history allows the fratricide to displace the original disobedience as the original sin. In any case, the notion of original sin resonates with Morton's exhortation that we take the precarious leap to act on issues such as global warming *even if* we are not strictly 'responsible' for it, *even if* it will not come about (*Ecology Without Nature*, p. 183).

But in Porphyre and Marble do appeare,
Might well haue hop'd to haue obtained it).

Giving the architectural buildings—whose ruins have been lamented for the previous 31 sonnets—precedence over literary works as that which might overcome the ravages of time gives the impression that 'le temps | Oeuvres & noms finablement atterre' ('time shall ruinate | Your works and names, and your last reliques marre'; Sonnet 7: 10–11). And yet, a hope that is set on the present remains. The poet finishes the sonnet sequence thus :

Ne laisse pas toutefois de sonner
Luth, qu'Apollon m'a bien daigné donner:
Car si le temps ta gloire ne desrobbe,
Vanter te peuls, quelque bas que tu sois,
D'avoir chanté le premier des François,
L'antique honneur du peuple à longue robbe. (9-14).

(Nath'les my Lute, whom *Phoebus* deigned to giue,
Cease not to sound these olde antiquities:
For it that time doo let thy glorie liue,
Well maist thou boast, how euer base thou bee,
That thou art first, which of thy Nation song
Th'olde honour of the people gowned long).

The use of the present tense in the tercets ('laisse' and 'peuls') focuses on the very act of writing, or enunciation at the present time and its value. Despite the seemingly inevitable assurance that time will bring everything to its natural end in death and destruction—the main theme of the sequence of sonnets—the last lines of the poem give the impression of deferring that end and giving precedence to the present provided it makes itself valuable.[41] That is, the lute is urged to keep playing only so long as it does it for a worthy cause: singing the 'antique honneur de people à la longue robbe'.

Indeed, the last sonnets highlight the value of the work of the present in face of the inevitable, fated end. The imagery used, moreover, again turns to the re-emergence of the natural landscape as the main theme, particularly in Sonnet 30, which compares Rome to a sown field:

41 Anderson, for example, argues against the pessimistic interpretations of Tucker. See 'La fin des Antiquitez de Rome: Vision pessimiste ou espoir de renouveau?'.

Comme le champ semé en verdure foisonne,
De verdure se hausle en tuyau verdissant,
Du tuyau se herisse en epic florissant,
Et comme en la saison le rustique moissone
Les undoyans cheveux du sillon blondissant,
Les met d'ordre en javelle, & du blé jaunissant
Sur le champ despouillé mille gerbes façonne :
Ainsi de peu à peu creut l'Empire Romain,
Tant qu'il fut despouillé par la Barbare main,
Qui ne laissa de luy que ces marques antiques,
Que chacun va pillant: comme on void le gleneur
Cheminant pas à pas recueillir les reliques
De ce qui va tumbant après le moissonneur.

(Like as the seeded field greene first showes,
Then from greene grasse into a stalke doth spring,
And from a stalke into an eare forth-growes,
Which eare the frutefull graine doth shortly bring;
And as in season due the husband mowes
The wauing lockes of those faire yellow heares,
Which bound in sheaues, and layd in comely rowes,
Vpon the naked fields in stackes he reares:
So grew the Romane Empire by degree,
Till that barbarian hands it quite did spill,
And left of it but these olde markes to see,
Of which all passersby doo somewhat pill:
As they which gleane, the reliques vse to gather,
Which th' husbandman behind him chanst to scater).

Within the pastoral scene, the humble work of the 'le rustique' and the 'gleneur' gain value as their work symbolizes the potential for continued conservation and growth. This image of the humble gleaner[42] anticipates Morrison's discussion of the poet as ragpicker who collects and reshapes what is discarded by others and whose work evokes both terror and beauty in the spectator.[43] Ironically, for Du Bellay's vision of Rome, the question of conservation was most related to Roman culture, its buildings and literature, but as a result of the destructive effects of war and time (in the poems'

42 On the image of the gleaner and the trope of humility see Tucker, *The Poet's Odyssey*, p. 19.
43 Morrison, *The Literature of Waste*, pp. 196–199.

logic the consequences of the city's original sin), what remains is nothing but rustic landscapes.

In the third sonnet, Du Bellay highlights the 'mondaine inconstance' ('worldly inconstancy') in which the apparently solid and stable cultural objects of an 'Eternal City' come to ruin while the fleeting and flowing river 'withstands' the ravages of time, precisely because it was always already in the process of change.[44] Just as the pastoral field of the eighteenth sonnet, where the land depicted changes over time from a pastoral landscape to a great civilized empire with palaces, only to return to its origin, here the river which was present before the erection of Rome's walls by Romulus is depicted as remaining after the fall of the empire—not as the same river, but as what remains ('reste') of Rome. And as Heraclitus's twelfth fragment might suggest, the return to the natural setting is not a return to the same origin, but rather to a new one: for '[a]s they step into the same rivers, different and (still) different waters flow upon them'.[45] The third sonnet's repetition of the word 'Rome', as critics have noted, points to the remains as something essentially different, thus emphasizing the change of the referent over time:[46]

Rome de Rome est le seul monument,
Et Rome Rome a vaincu seulement.
Le Tibre seul, qui vers la mer s'enfuit,
Reste de Rome. O mondaine inconstance!
Ce qui est ferme, est par le temps détruit,
Et ce qui fuit, au temps fait résistance. (9–14).

(Rome now of Rome is th'onely funeral,
Ans onely Rome of Rome hath victorie;
Ne ought saue Tyber hastning to his fall
Remaines of all: O worlds inconstancie.
That which is firme doth flit and fall away,
And that is flitting doth abide and stay).

44 Haldane, like others emphasize the paradox of this image where the river that remains from Rome (and thus represents the city) flees itself. See 'Et Rome Rome a vaincu seulement', p. 476.
45 Even the stones (*pierres*) of the opening quatrain of the eighteenth sonnet return at as 'successeur de Pierre', and the shepheards (*pasteurs*) return merely in name ('sous nom de pasteur') thus clearly functioning in a different, figurative way. It is worth noting that the return to the origin privileges the Christian tradition over the pagan one as the pagan monuments and pastors are transformed into Christian symbols.
46 As Melehy shows, du Belay had translated passages from Ovid that transmitted Heraclitus's image of the stream streaming, in 'Du Bellay's Time in Rome', p. 12.

Here the destructive effects of time on human cultural artefacts is highlighted insofar as the hubris of a culture that would build an 'eternal city' is punished by destruction. Unexpectedly, the sonnet also explicitly identifies the Tiber river—nature—as the sole recognizable remnant of that great city.[47]

Toward a Conclusion

Du Bellay's images, though not always literal representations of nature, anticipate concrete examples of the unanticipated re-emergence of nature from the ruined remains of modern culture. We are entering into a peculiar kind of ruin resembling modern toxic land dumps or post-military landscapes. Because of their restricted access, these spaces then 'return' to being beautiful natural landscapes that can be turned into nature reserves.[48] The sonnets suggest a constantly shifting relation between nature and culture in which natural elements such as the Tiber river are depicted already as being parts of the city as much as the city is forever to be evoked by the river.[49] The contrast of imperial pride and wickedness with celestial power

47 The river returns in the opening of the second sequence of sonnets on the theme of Rome's ruins, *Songe*. There, the poet is addressed by a 'Demon [...] | Dessus le bord du grand fleuve de Rome', pp. 5–6 (a Ghost [...] | On that great riuers banck, that runnes by *Rome*). However, in contrast to its figuration in the *Antiquitez*, here resistance to the destructiveness of time is not ascribed to the river, but rather to God: 'Puis que Dieu seul au temps fait ressistence, | N'espere rien qu'en la divinité', pp. 13–14 ('Sith onely God surmounts all times decay, | In God alone my confidence do stay').

48 The description of the United States' 'most ironic nature park', Denver's Rocky Mountain Arsenal in *Uncommon Ground: Rethinking the Human Place in Nature* is instructive here in its 'juxtaposition of toxicity and wilderness', as is the management of post-military landscapes into areas of 'exceptional nature conservation value' in the United Kingdom. See also Desilvey, *Curated Decay*, p. 88.

49 It is worth noting that from beginning to end, the Tiber itself cannot be freed from its association to Rome—even before the establishment of the city, the river plays a crucial role in the foundational narrative of Romulus and Remus. According to the legend, Amulius overthrew Numitur, his brother and King of Alba Longa, and killed his sons. He also forced his wife Rhea Silva to become a Vestal Virgin to prevent her from having other sons who might oppose him in the future. And yet, as a result of being either raped or seduced by Mars, Rhea became pregnant. Rather than killing Rhea for 'breaking' her chastity vows, Amulius ordered that her infant sons, Romulus and Remus, be drowned in the Tiber River. But instead, they were placed in a basket which under the care of the river deity Tiberinus floated down the river and came to rest at the site of the future city near the *Ficus ruminalis*. Interestingly, the Christian narrative that Du Bellay superimposes—from both Genesis and Revelation—starts and ends with a river that is both the same and different, and which is linked to both the garden and the city. See Genesis 2. 10–14; Revelation 22. 1–3.

and justice enacted through the destructiveness of civil war and time gives moral significance to the 'universal flux'.

Though his primary focus is, for sure, anthropocentric, Du Bellay represents nature's resistance to its inevitable destruction by including a temporal aspect in addition to the spatial one. The structure of the sonnet sequence, moreover, creates a movement that reflects the re-emergence of nature among the ruins of the ancient city: from the pomp of palaces in the opening sonnet *Au Roy* and the prideful display of architectural achievement including contemporary French buildings ('Sainct-Germain', 'Fontainebleau'),[50] and ancient structures (the seven wonders of the world cited in Sonnet 2), we are led to the ruins that are left behind, the remaining architecture described as a body in ashes (Sonnet 5), and buildings whose ephemeral grandeur only 'au temps pour un temps facent guerre' ('do for a time make warre | Gainst time'; Sonnet 7). Rome's military accomplishments are praised (Sonnet 8) only to highlight that it was precisely its pride and origins in war that would be its undoing (Sonnets 10, 23, 24, and 31). As the ruins are slowly depicted in the sequence, the landscape becomes quickly appropriated by nonhuman nature (symbolic animals in Sonnets 14 and 17; elements in Sonnets 16 and 20; pastoral landscapes in Sonnets 18 and 30). The architectural design of the sonnet sequence seems to point to a triumphal renascent nature re-emerging from, and appropriating the ruins of a prideful and war-torn civilization. The intersection of these themes as presented in the *Antiquitez* invites further questions: how can we articulate the goals of conservation beyond utilitarian or Neoplatonic paradigms where nature is preserved insofar as it is useful or as a relic of the past?[51] Moreover, the poems suggest that, in a way, the question of ecological conservation can be interpreted temporally, as the question both of nature's resistance against time and of our own lack of authenticity, since we temporally displace the moment of death (Nature's and our own). That is, we are ever more aware of the destructive future[52] consequences our actions will have on the

50 As McGowan points out, the château of Fontainbleau was largely shaped by influences from Rome, and housed collections of Roman art (*The Vision of Rome*, pp. 173, p. 346).

51 See Sagoff's criticism of new conservationists, 'Kareiva among them, [who] have argued that nature might better be preserved for its utilitarian value, not as a commodity and input to material production but rather for the ways in which ecosystems, on the model of Eden, provide useful services to human societies' ('What Is the Nature We Seek to Save?', p. 6). Rabelais's description of Gaster and the Gastrolatres in the *Fourth Book* might provide fertile ground to analyze the Edenistic/utilitarian argument. See Krause, 'Idle Works in Rabelais' Quart Livre: The Case of the Gastrolatres'.

52 Although it might also be accurate to say with Morton the far from being imminent, the ecological catastrophe '*has already taken place*', p. 28.

environment and our own species' chances for survival, while at the same time we refuse the fact that those consequences have anything to do with us.

Although contemplating death was not uncommon in the Renaissance, accepting it as natural was still difficult.[53] Our reactions to the ruins of the past and the image of the potential ruinous future of our environment reveals an inauthentic attitude toward death that Heidegger articulates as our inability to grasp Nature's Being in its totality—especially as we continue to evade its death and our complicity in it through the everydayness of a phrase like, 'One of these days one will die too, in the end; but right now it has nothing to do with us'.[54] Du Bellay's ambivalence regarding Rome's ruins and his own mortality (both literary and literally) in the *Antiquitez* challenges us to find beauty even in our decaying landscapes. It calls us to recognize our place in our current situation (*situ*—site and decay)[55] and, like the humble gleaner, to never stop yielding new value out of the discarded material before us.

Works Cited

Anderson, Lynn, 'La Fin Des Antiquitez De Rome: Vision Pessimiste Ou Espoir De Renouveau?' *Romance Quarterly* 56.2 (2009), 114–126.

Aristotle, 'Physics', in *The Basic Works of Aristotle*, ed. by Richard McKeon, (New York: Random House, 1941).

Belleau, Remy, *Oeuvres Complètes de Remy Belleau*, ed. by A. Gouverneur, Vol. III. (Paris: A. Gouverneur, 1777).

Desilvey, Caitlin, *Curated Decay: Heritage Beyond Saving* (Minneapolis: University of Minnesota Press, 2017).

Du Bellay, Joachim, *Les Regrets et autres oeuvres poëtiques suivis des Antiquitez de Rome, plus un Songe ou Vision sur le mesme subject* (Geneva: Librarie Droz, 1974).

———, *Joachim du Bellay: 'The Regrets' with 'The Antiquities of Rome,' Three Latin Elegies, and 'The Defense and Enrichment of the French Language'*, ed. and trans. by Richard Helgerson (Philadelphia: University of Pennsylvania Press, 2006).

53 For example, when Montaigne considers the chances of dying a 'natural death,' he notes that 'Nous l'apellons seule naturelle, comme si c'estoit contre nature de voir un homme se rompre le col d'une cheute, s'estoufer d'un naufrage, se laisser surprendre à la peste' (*Essais*, I.LVII. 326a) ('We call it alone natural, as if it were contrary to nature to see a man break his neck by a fall, be drowned in a shipwreck, or be snatched away by the plague').

54 Heidegger, *Being and Time*, p. 297.

55 See Morton's description of dark ecology in *Ecology Without Nature*, pp. 195–196.

Egan, Gabriel, *Green Shakespeare: From Ecopolitics to Ecocriticism* (New York: Routledge, 2006).

Fudge, Erica, *Renaissance Beasts* (Urbana: University of Illinois Press, 2004). EPUB file.

Gadoffre, Gilbert, *Du Bellay et le sacré* (Paris: Gallimard, 1978).

Glasser, Richard, *Time in French Life and Thought*, trans. by C.G. Pearson (New Jersey: Rowman & Littlefield, 1972).

Gouwens, Kenneth, 'What Posthumanism Isn't', in *Renaissance Posthumanism*, ed. by Joseph Campana and Scott Maisano (New York: Fordham University Press, 2016), pp. 37–63.

Haldane, Michael, '"Et Rome Rome a Vaincu Seulement": Du Bellay's Sonnet III of the *Antiquitez De Rome*', *Neophilogus* 97 (2013), 465–480.

Harrison, Robert, *Forests: Shadows of Civilization* (Chicago: University of Chicago Press, 1992). EPUB file.

Heidegger, Martin, *Being and Time*, trans. by John Macquarrie and Edward Robinson (New York: Harper & Row, 2008).

Heraclitus, *Heraclitus: Fragments*, Phoenix Presocratic Series, (Toronto: University of Toronto Press, 1987).

Holy Bible, New International Version (Colorado Springs: International Bible Society, 1973).

Hui, Andrew, *The Poetics of Ruins in Renaissance Literature*, (New York: Fordham University Press, 2016).

Krause, Virginia, 'Idle Works in Rabelais' Quart Livre: The Case of the Gastrolatres', *The Sixteenth Century Journal* 30.1 (1999), 47–60.

Lewis, Simon L., and Mark A. Maslin, 'Defining the Anthropocene', *Nature* 519 (12 March 2015), 171–180.

McGowan, Margaret, *The Vision of Rome in Late Renaissance France* (New Haven: Yale University Press, 2000).

Melehy, Hassan, 'Du Bellay's Time in Rome: The Antiquitez', *French Forum* 26.2 (2001), 1–22.

Merrill Ingram, Annie, *Coming into Contact: Explorations in Ecocritical Theory and Practice*, ed. by Annie Merrill Ingram, (Athens: University of Georgia Press, 2007).

Montaigne, Michel de, *Les Essais, livres I-III* (Paris: Presses Universitaires de France, 2002).

———, *The Complete Essays of Montaigne*, trans. by Donald M. Frame (Stanford: Stanford University Press, 1957).

Morrison, Susan Signe, *The Literature of Waste: Material Ecopoetics and Ethical Matter* (New York: Palgrave MacMillan, 2015).

Morton, Timothy, *Ecology Without Nature: Rethinking Environmental Aesthetics* (Cambridge: Harvard University Press, 2007).

Rabelais, François, *Le tiers livre* (Torino: Le Livre de Poche, 1994).
———, *Gargantua and Pantagruel*, trans. by Burton Raffel (New York: W.W. Norton & Company, 1990).
Rickards, Lauren A., 'Metaphor and the Anthropocene: Presenting Humans as a Geological Force', *Geographical Research* 53.3 (2015), 280–287.
Russell, Daniel, 'Du Bellay's Emblematic Vision of Rome', *Yale French Studies* 47 Image and Symbol in the Renaissance (1972), 98–109.
Sagoff, Mark, 'A Theology for Ecomodernism', *Breakthrough Journal* 5 (2015), https://thebreakthrough.org/journal/issue-5/a-theology-for-ecomodernism. Accessed 27 September 2019.
Spencer, Edmund, 'Ruines of Rome: by Bellay', in *The Works of Edmund Spencer: The Minor Poems Volume 2* (Baltimore: The Johns Hopkins Press, 1947), pp. 141–154.
Tucker, George Hugo, *The Poet's Odyssey: Joachim Du Bellay and the 'Antiquitez De Rome'* (Oxford: Clarendon Press, 1990).
Usher, Phillip John, 'Untranslating the Anthropocene', *diacritics* 44.3 (2016), 56–77.
Virgil, *The Aeneid*, trans. by Robert Fitzgerald (New York: Random House, 1990).
Yandell, Cathy, *Carpe Corpus: Time and Gender in Early Modern France* (Newark: University of Delaware Press, 2000).

About the Author

Victor Hugo Velázquez (Associate Professor of Spanish and French, and Chair of Modern Languages at Biola University) has published research on the intersection of French Renaissance literature, ecocriticism and Christian thought (*Journal of Christianity and World Languages*), intertextuality in Michel de Montaigne (*Utah Foreign Language Review, Inhabited by Stories*), as well as post-secularism in Samuel Beckett's *Waiting for Godot* (under review with *Christianity and Literature*). He has also presented his research on the complexity of the representation of nature-culture in Joachim du Bellay's *Antiquitez de Rome*, the concept of freedom and sovereignty in Étienne de la Boétie and Molière, as well as history and narrative in Miguel de Cervantes.

Part 3

Groundings

9. An Inconvenient Bodin: Latour and the Treasure Seekers

Oumelbanine Nina Zhiri

Abstract

The admirers of Jean Bodin's political philosophy might be surprised by his *Demonomanie des sorciers* and his forceful attempt to prove the reality of witchcraft. This opposition between the enlightened modern and the superstitious premodern makes his thought a prime example to confront to the theory of modernity proposed by Bruno Latour. This essay attempts such an exploration, and focuses on narratives of treasure seeking in Bodin's text, to understand the notion of nature that they bespeak, a nature entirely worked through by demons. Looking at Bodin as a premodern also allows us to complicate Latour's account by highlighting what the resurgence of thinking about witchcraft in late sixteenth-century Europe reveals about a larger argument about Nature, and the ways in which humans should deal with it.

Keywords: Jean Bodin, Bruno Latour, Nature, witchcraft, treasure hunting

In his treatise on witchcraft *De la démonomanie des sorciers* (1580), Jean Bodin (1530–1596) tells the story of a friend who, with a group of companions, went looking for buried enchanted treasures in Lyon. Their quest was unsuccessful as the spirits guarding the treasure scared them off with a horrific scream. Such a story might surprise those more used to thinking of Jean Bodin as a precursor of the rational Enlightenment. How can we make sense of this story and of the worldview it bespeaks? The present study will suggest that the work of Bruno Latour can help us answer such a question, and that, in return, Bodin's text can present a fresh perspective on Latour.

Goul, P. and P.J. Usher (eds.), *Early Modern Écologies. Beyond English Ecocriticism.*
Taylor & Francis Group, 2020
DOI 10.5117/9789462985971_CH09

The Early Modern Bifurcation of Nature

For those who study the early modern period, the work of Bruno Latour—and that of the many scholars who are building on his insights—is of special interest, as Louisa Mackenzie and others have argued.[1] Latour's works, including *We Have Never Been Modern* and *Pandora's Hope,* analyse early modern 'revolutions', the Renaissance, and the Scientific Revolution, in order to question the validity of our collective construction of modernity.[2] Conversely, the study of early modern culture can help shape our reading of Latour. According to Latour, the invention of Western modernity is rooted in a representation of nature which itself based on an originary bifurcation, a concept borrowed from Alfred North Whitehead, who explored how early modern philosophers and their followers created a clear and ever-widening separation between 'what is in the mind, and what is in nature'.[3] Latour's analysis points to the way in which modernity has not in reality eliminated mediation and mediators, but works at covering up their processes, and thus produces divides between stark binary opposites, such as Nature and Culture, or subject and object.

The concept of nature, central to the construct of modernity, founds the radical distinction between the West and all other cultures (including premodern Europe): 'Nous, les Occidentaux, ne pouvons être une culture parmi d'autres, puisque nous mobilisons aussi la nature. Non pas, comme le font les autres sociétés, une image ou une representation symbolique de la nature, mais la nature telle qu'elle est, du moins telle que les sciences la connaissent, sciences qui demeurent en retrait, inétudiables, inétudiées'.[4] ('We Westerners cannot be one culture among others, since we also mobilize Nature. We do not mobilize an image or a symbolic representation of Nature, the way the other societies do, but Nature as it is, or at least as it is known to the sciences—which remain

1 Mackenzie, 'It's a Queer Thing'.

2 Latour, *Nous n'avons jamais été modernes / We Have Never Been Modern*; *Pandoras's Hope: Essays on the Reality of Science Studies.*

3 Whitehead, *The Concept of Nature*, p. 20. Whitehead expands on this idea when he describes the 'bifurcation of nature into two systems of reality, which, in so far as they are real, are real in different senses. [...] into two divisions, namely into the nature apprehended in awareness and the nature which is the cause of the awareness. The nature which is in fact apprehended in awareness holds within it the greenness of the trees, the song of the birds, the warmth of the sun, the hardness of the chairs, and the feel of the velvet. The nature which is the cause of awareness is the conjectured system of molecules and electrons which so affects the mind as to produce the awareness of apparent nature' (p. 21).

4 Latour, *Nous n'avons jamais été modernes*, p. 132.

in the background, unstudied, unstudiable').[5] There is another crucial element of the distinction between the Moderns and the non-Moderns: 'Les autres cultures frappèrent toujours les modernes par l'aspect diffus de leurs forces actives ou spirituelles. Nulle part elles ne mettaient en jeu de pures matières, de pures forces mécaniques. Les esprits et les agents, les dieux et les ancêtres se mêlaient en tous points'.[6] ('Moderns were always struck by the diffuse aspect of active or spiritual forces in other so-called premodern cultures. Nowhere were pure matters, pure mechanical forces, put into play. Spirits and agents, gods and ancestors, were blended in at every point').[7] Jean Bodin's treasure seekers inhabit this premodern world, in which supernatural forces intervene, where spirits, humans, and objects mix and fight each other, and in which nature and culture intermingle. By analyzing how Bodin pictures their activity and contrasts it with other underground enterprises, such as mining, this study will examine the ontology and epistemology that sustains it. It will moreover highlight the complexities of early modern notions of nature, and complicate Latour's historical narrative.

Witchcraft, Magic and Treasure Seeking

Jean Bodin is a particularly interesting author to re-read when reflecting on the coming of modernity. His influential and erudite books explore the fields of law, history, and nature, as well as their multiple interconnections. His work illuminates how paradoxical, contradictory, and messy the path was that led to the dominance of the construct of modernity in European culture. Best known for a seminal text of political philosophy, *Les Six Livres de la République* (1576), when Bodin writes about government and religion, and in his own personal engagement with the more tolerant faction during the Wars of Religion, he appears to modern readers to have been one of the most enlightened minds of his time. This reputation is potentially sullied by *De la démonomanie des sorciers* (1580), in which he aimed at proving the reality of witchcraft. He stood in stark opposition to some of his contemporaries,

5 Latour, *We Have Never been Modern*, p. 97. Elsewhere, he adds these important precisions on the modern concept of nature, which is 'not a thing, a domain, a realm, an ontological territory. It is [...] a way of organizing the division (what Alfred North Whitehead has called the Bifurcation) between appearances and reality, subjectivity and objectivity, history and immutability' ('An Attempt at a "Compositionist Manifesto"', p. 476).

6 Latour, *Nous n'avons jamais été modernes*, pp. 175–176.

7 Latour, *We Have Never been Modern*, p. 128.

like Johann Weyer (1515–1588), who believed that when witches confessed to participating in the sabbath or to making deals with the devil, they were the victims of illusions or hallucinations. On the contrary, Bodin belonged to the 'realist' school that was convinced of the actuality of the sabbath and of witchcraft.[8] In his view, witchcraft presented an extreme danger to society, and to humanity even. Thus Bodin vociferously advocated for witch-hunting, and for the harshest punishments for those found guilty; as a result, he was considered an extremist even among his contemporaries in the law profession. His effort was part of a resurgence of witch-hunting in the late sixteenth and early seventeenth centuries that was accompanied by a flurry of new publications, among which his *Démonomanie* stands out for both its erudition and for its commitment to this cause.

In the words of Ann Blair, Bodin can thus be said to be somewhat 'Janus-faced, torn between the modern insights of the *République* and the superstitions of the *Démonomanie*', a dichotomy undeniably based on the prevalent narrative of modernity as the path from a time of ignorance to one of enlightened rationality.[9] Recent scholarship has however moved from seeing this text as a 'monstrous excrescence' in Bodin's work, and towards a better understanding of how it fits in the overall context of his œuvre.[10] Putting this debate aside, this study will focus on a few pages devoted to enchanted buried treasures and their seekers. This debate is nevertheless relevant because it radically precludes understanding Bodin's intervention on treasure hunting as just another episode in the grand narrative of modernity, and stops short of concluding that Bodin was proposing a more rational way to deal with nature as an alternative to treasure seeking.

Jean Bodin's analysis of treasure seeking stands out for the period. Contemporaries rarely analysed this activity, and most early modern accounts simply document it without offering any examination. Bodin, through multiple textual strategies (including parallels, narratives, and characters), offers an original reflection on the meaning of the practice of treasure seekers, which will allow us to explore his views on the relationship of humans to objects and matter. The story of Jean Bodin's friend in Lyon was far from exceptional in his time. According to a recent study, 'it was common knowledge in Old European culture, that is, the culture before the onset of the Industrial Revolution that spirits guarded treasures and sometimes

8 See Krause, 'Listening to Witches: Bodin's Use of Confession in *De la Démonomanie des Sorciers*'.

9 Blair, *The Theater of Nature. Jean Bodin and Renaissance Science*, p. 12.

10 Jacques-Chaquin, 'La "Démonomanie des sorciers", une lecture politique de la sorcellerie', p. 44.

gave their riches to men'.[11] Closely linked to the supernatural, treasure seeking was a specifically early modern phenomenon, at least in Europe.[12] The search for hidden treasures with the help of magic was attested all over premodern Europe.[13] This activity was illegal, as were other forms of magic, and European court records on this activity are numerous, thus providing much rich material for historians.

Scholars have mostly focused on the social and legal aspects of treasure hunting, only hinting at epistemological connections, including the idea that this activity contributed to the birth of archaeology.[14] Dillinger also proposes that treasure hunting illustrates the slow transformation from agricultural to bourgeois capitalistic society, seeing it as a new form of economic initiative taken by urban dwellers, either craftsmen or people from the middle strata of society, looking for social mobility. Their entrepreneurial and profit-seeking initiative had already taken them at least some way from the static agricultural society based on the concept of the limited good, although still far from the early modern large-scale financial transactions or international trade that were beginning to take place. They are thus transitional figures of a society in the process of transforming toward a market capitalist economy. Jean Bodin's interest in economics might have made him take notice of this widespread activity and could explain why he devoted several pages to it. His account is a rarity, since treasure-hunting is only tangentially related to witchcraft and was thus barely mentioned in the abundant demonological literature. Seekers resorted to magic and were thus subject to legal prosecution, but they usually did not turn to Satan and his minions, and, as a result, did not risk the same harsh pursuits and punishments as witchcraft. According to Dillinger,

> All kinds of spirits and a huge arsenal of magical items figure in the trial records of treasure hunters as well as in treasure lore, but witches and

11 Dillinger, *Magical Treasure Hunting in Europe and North America, A History*, p. 1. On the phenomenon in Europe, see Sallmann, *Chercheurs de trésors et jeteuses de sort: la quête du surnaturel à Naples au seizième siècle* and Bercé, *À la recherche des trésors cachés du XVIème siècle à nos jours.*

12 Dillinger, *Magical Treasure Hunting in Europe and North America*, p. 8: 'most treasure hunts took place in the early modern period'.

13 It was also prevalent in North Africa and the Middle East, where it probably harked back to much earlier times, even ancient Egypt. For a recent study on the phenomenon in Egypt, see Braun, *Treasure Hunting and Grave Robbery in Islamic Egypt. Textual Evidence and Social Context.* There has been to my knowledge no study taking into account the transcultural aspect of this phenomenon.

14 Dillinger, *Magical Treasure Hunting in Europe and North America*, p. 20: 'Treasure hunting was one of the earliest beginnings of archaeology. The first more or less organized excavations in the ruins of antiquity took place because people were looking for treasure'.

> witchcraft play next to no role. [...] Treasure hunting, very like other forms of folk magic, such as simple charms against common ailments or the host of practices supposed to protect crops and livestock, was never seen as a typical activity of witches. Learned demonology was not really interested in the treasure hunt, even though it was clearly a magical activity.[15]

Narratives of Treasure Seeking

At first glance, one can legitimately wonder why Bodin would devote the best part of a chapter of a book concerned chiefly with witchcraft to treasure seekers. I will suggest that his treatment of the topic exceeds the limits of demonological thinking. Even though the author, as a jurist, had a keen interest in the law, his view of the topic is not juridical, and offers far more in terms of an epistemological understanding of the treasure seekers' activity and of human engagement with the realm of matter, especially when this matter is located under the ground.

Jean Bodin takes up the subject of the treasures seekers in Chapter III of Book III of the *Démonomanie,* titled 'Si les sorciers peuvent avoir par leur mestier la faveur des personnes, la beauté, les plaisirs, les honneurs, les richesses, & les sciences, & donner fertilité' ('On whether witches can by their craft obtain the favour of people, beauty, pleasures, honors, riches, and sciences, and offspring').[16] This chapter is devoted to exploring, and forcefully denying, the efficacy of satanic help or magical practices in obtaining desirable worldly goods. On the contrary, it insists that those who seek the assistance of Satan to obtain beauty, riches, honour, and knowledge are in fact ugly, poor, despised, and ignorant. This, ostensibly, is the main point that Bodin wants to make in this chapter. However, among these desirable possessions and attributes, the only aspect that he treats at some length is the theme of riches. Three pages suffice for all the other issues. The remaining six focus almost entirely on the issue of buried enchanted treasures. In his rapid review of the other desirable attributes (beauty, honour, favour), Bodin very emphatically makes the single main point already mentioned: that all good things come from God, that Satan does not and cannot bestow them to his devotees, and neither can magic, since God's will, and the laws of nature, cannot be circumvented.

Bodin's account freely combines quotations, personal anecdotes, conversations with friends and colleagues, in a kind of *bricolage* that reminds

15 Dillinger, *Magical Treasure Hunting in Europe and North America*, p. 136.

16 Bodin, *De la Démonomanie des sorciers*, p. 305.

the reader that Renaissance literature delighted in developing 'hybrid and uncanonical forms', and 'mixing heterogeneous materials'.[17] In the Lyon story, the hunters hear a terrifying scream: 'You thieves!' Bodin concludes the story as follows: 'Ainsi void on que les malings esprits ne veulent pas, ou pour mieux dire, que Dieu ne souffre pas que personne par tels moyens puisse enrichir' ('One can thus see that the evil spirits don't want or, to say better, that God does not allow anybody to get rich this way'),[18] noting that Scripture gives riches the name of blessings (*benedictions*) because they are gifts from God. Then he presents more arguments to make his case, using different approaches to explore and illustrate them. The treasure is not a metaphor but a theoretical construct through which Bodin discusses ways in which humans connect with things and matter, and deal with the nonhuman. This is not to suggest that Bodin did not believe in the reality of the treasures, but merely to propose that he used them for a larger reflection, which could explain why he took on this unusual topic.

This reflection is deployed through a series of stories, either told to the author by friends or borrowed from books. This narrative form, so prevalent in this text, is not simply a way for Bodin to acknowledge or retrace his sources; it allows him to define, explore, and complicate the meanings of treasure and of treasure hunting. Beyond mere illustration, the tales map the connection between the hunters and their goal, and ultimately between the human seeker and the natural realm; they help us understand how treasure seeking is a network of humans, spirits, objects, and (maybe) Satan and his agents. To use one of Latour's concepts, treasure seeking is a hybrid, a quasi-object. As Graham Harman has summarized, 'to follow a quasi-object is to trace a network'.[19] Bodin's storytelling allows him to change elements of the networks, to explore how actants connect and interact, and thus to deploy different meanings of treasure seeking. Through the stories describing the hunt for enchanted treasures, Bodin will get to the point when he can engage in a larger reflection on how people deal, beyond these specific things, with other matters that are located below the Earth's surface; the treasure will then figure as just one modality among others of the hidden subterranean object.

He starts by stating a widely known fact: 'Quant aux richesses, on sçait assez qu'il y a de grands tresors cachez, & que Satan n'ignore point les lieux où ils sont' ('Concerning riches, it is well known that there are great

17 Blair, *The Theater of Nature. Jean Bodin and Renaissance Science*, p. 15.

18 Bodin, *De la Démonomanie des sorciers*, p. 310.

19 Harman, *Prince of Networks. Bruno Latour and Metaphysics*, p. 64.

hidden treasures, and that Satan does not ignore where they are found').[20] However the devil will not be able to bestow the treasures even to his faithful disciples, because 'Dieu ne le veut pas, et [...] le Diable n'a pas la puissance' ('because God does not want it, and the devil does not have the power').[21] The enchanted treasure according to Bodin is and will remain unattainable, although it commands the pursuit of numerous actants. The anecdotes explore different ways in which this impossibility becomes inescapable.

In one story, an acquaintance of the author, a physician from Toulouse named Oger Ferrier, rented a haunted house. He sought the help of a Portuguese student 'qui faisoit voir sur l'ongle d'un jeune enfant les choses cachees' ('could show on the nail of a young child hidden things'), and, thanks to a small girl, uncovered the location of a buried treasure.[22] These traits are common in contemporary stories of treasure hunting: someone who masters magic, often a foreigner, serves as a guide; children are involved, seen as being able to divine the places where treasures were hidden.[23] The physician dug to find the treasure, but a mighty wind caused the collapse on his house of part of the neighbouring building. The next day, the Portuguese student explained to Oger Ferrier that the spirit had left and taken the treasure with him. In his friend's house, Jean Bodin witnessed the material traces of the mayhem that occurred. In this story, neither Satan nor witchcraft are directly involved, only more benign magic and spirits, which might explain why the punishment is only loss of property. However, as was explicated by the Portuguese student, the demons were involved: they usually are when violent winds blow in times other than the natural seasons, as Bodin himself would later underline in the *Universae naturae theatrum* (1596).[24]

The lives of the treasure seekers are also spared in other stories, such as that of the acquaintance of Bodin's from Lyon who, having learned his lesson, vowed to renounce treasure hunting. Another hunter, however, who used the help of magic to find a treasure in Arcueil near Paris, is left crippled for life.[25] Other tales show that treasure hunting becomes a much more perilous activity when one elicits help beyond usual magic, and looks for the assistance of the devil. One, borrowed from Phillip Melanchthon (1497–1560) recalls the death of ten treasure seekers 'lors qu'ils fossoyoyent pour trouver les thresors que Sathan leur avoit enseignez' ('while they were

20 Bodin, *De la Démonomanie des sorciers*, p. 308.
21 Ibid., p. 308.
22 Ibid., p. 309.
23 Dillinger, *Magical Treasure Hunting in Europe and North America*, p. 161; p. 153.
24 Céard, *La Nature et les prodigies*, pp. 358–359.
25 Bodin, *De la Démonomanie des sorciers*, p. 310.

digging to find the treasures that Satan had informed them about').[26] In Nuremberg a priest, 'ayant trouvé un thresor à l'aide de Sathan' ('having found a treasure with Satan's help') was on the point of opening a chest when the house collapsed upon him.[27]

A New Ontology of the Subterranean

Beyond the simple moral lesson of the stories, their evolution and the type of emphasis put on different elements help Bodin construct more complex meanings, mobilize different actants and describe new modes of relations, further opening the reflection. An essential aspect of treasures in Bodin's text is that they are buried. Early on, Bodin underlines that the Bible condemns the very act of burying treasures, of putting them under the surface of the earth,[28] since doing so not only reveals avarice, but also puts treasures—even when not dishonestly acquired—under the dominion of demons that rule the underground.[29] This signification is further broadened given that the subterranean world during the sixteenth century was the object of important new discourses that bring into focus the question of how one should deal with the underground in order to acquire wealth.[30] Bodin's account connects with these new developments by quoting the most influential author in this momentous change:

> Et Georgius Agricola au livre qu'il a fait des Esprits subterrains, escript qu'à Aneberg en la mine nommee Couronne de Roze, un esprit en forme de cheval tua douze hommes, tellement qu'il fit quitter la mine pleine d'argent, que les Sorciers avaient trouvé à l'ayde de Sathan.[31]
>
> (And Georgius Agricola, in the book he made about the subterranean spirits, writes that in Aneberg, in the mine called Crown of Rose, a spirit in the guise of a horse killed twelve men, with the result that it made

26 Ibid., p. 309.
27 Ibid., p. 310.
28 Ibid., p. 309.
29 Dillinger, *Magical Treasure Hunting in Europe and North America*, p. 62, notes that 'the subterranean realm was often seen as the realm of demons'.
30 On mining and the subterranean worlds in the early modern period, see Usher, *Exterranean: Extraction in the Humanist Anthropocene*.
31 Bodin, *De la Démonomanie des sorciers*, p. 309. See Agricola, *De animantibus subterraneis liber*, p. 77.

people to desert the mine although it was full of silver, that witches had discovered with the help of Satan).

It is worth noting that Agricola's original anecdote mentions neither Satan nor witches, but merely illustrates that demons are found underground. In Bodin's retelling, the death of the miners is not simply an unfortunate event: it is also a sign of God's wrath at their recourse to witchcraft. Furthermore, this quote, while ostensibly making the same point as the other anecdotes, in fact indicates a serious shift in focus, and adds to the text crucial new meanings and new ways of looking at the subterranean. The people who met such a brutal end in Aneberg were not looking for a treasure but for a silver mine. Talking about contemporary texts concerning treasure hunting, Dillinger concludes that 'it is safe to say that all of them agreed that there was nothing natural about treasure. Treasure was an artefact or consisted of artefacts. It was clearly no natural resource akin to the gold found in a mine'.[32] Miners did, however, sometimes look for valuable minerals with the help of magic, including dowsing and incantations. Georgius Agricola was aware of these practices, and indeed wrote against them. In the Aneberg case as retold by Bodin, miners went beyond those magical and divinatory practices, and resorted to witchcraft and sought the help of the devil, making their activity as reprehensible as the worst cases of treasure seeking described by Bodin.

Another story helps the author draw the line more precisely between acceptable and illicit ways of seeking the kind of wealth that is located underground:

> Et qui plus est, les souffleurs Alchemistes pour la pluspart, voyans qu'ils ne peuvent venir à bout de la pierre Philosophale, demandent conseil aux esprits, qu'ils appellent familiers. Mais j'ay sçeu de Constantin, estimé entre les plus sçavans en la Pyrotechnie, & art metallique, qui soit en France, & qui est assez cogneu en ce royaume, que ses campaignons ayant long temps soufflé sans aucune apparence de proffit, demanderent conseil au Diable s'ils faisoient bien, & s'ils en viendroient à bout. Il fit response en un mot; 'Travaillez.' Les Souflleurs bien aises continuerent, & soufflerent si bien qu'ils multiplierent tout en rien, & souffleroient encores n'eust esté que Constantin leur dist, que Sathan rendoit tousjours les oracles à double sens, & que ce mot 'travaillez' vouloit dire, qu'il failloit quitter l'Alchemie & s'employer au travail, & honneste exercice de quelque bonne science

32 Dillinger, *Magical Treasure Hunting in Europe and North America*, p. 3.

> pour gaigner sa vie, & que c'estoit une pure follie de penser contrefaire l'or en si peu de temps, veu que nature y employe mille ans.[33]
>
> (Moreover, Alchemist blowers for the most part, seeing that they cannot obtain the philosopher's stone, ask the advice of spirits. But I learned from Constantin, considered to be among the most knowledgeable in Pyrotechny, and metallic art, in France, and is rather well known in this kingdom, that his companions having blown for a long time without any appearance of profit, ask the devil to tell them if they were doing well, and if they will obtain what they were looking for. He said only one word: 'Work!' They were glad, and kept blowing, without obtaining anything, and would still be if Constantin had not told them that Satan's oracles always had a double meaning, and that this word, 'Work', meant that they should leave alchemy and devote themselves to work, to some honest exercise of some good science to make a living, and that it was pure folly to think of counterfeiting gold in such little time, since nature takes a thousand years).

Thanks to Bodin's informant (who is, like Georgius Agricola, a specialist of metallurgy and of subterranean matter),[34] this tale connects the alchemist's activity to the miner's. If the latter's goal is to excavate precious metals, the former seeks to create gold, and to emulate the underground work of nature. As in the Aneberg anecdote, the protagonists are misguided enough to ask Satan for help in their endeavours. Constantin himself intervenes in the story when he counsels the alchemists, and interprets the laconic response given by Satan: 'Work'. The companions understood it as meaning that they had to keep looking for the philosopher's stone to be able to make gold. Constantin's explanation is different: they had to abandon alchemy altogether and to devote themselves to work and to the honest exercise of some good science. One could suspect that he had in mind the newly formalized science of metallurgy, which could help them find natural gold through the hard work of mining.

The text has thus moved from the dubious effort of treasure seeking to the honest craft of mining. Bodin, through a series of narratives, has proposed different configurations for a network of actants, and their evolution deserves a closer analysis. First, one looks for a treasure, an artefact made by humans and buried underground, which is guarded by spirits that

33 Bodin, *De la Démonomanie des sorciers*, p. 311.

34 This character remains unidentified.

the seeker has to circumvent by the (misguided) use of magic. He must often work with others, companions such as a specialist of magic (like the Portuguese student) or sometimes a child. Another network is even more dangerous, because it links humans and Satan, usually through the assistance of witches. This can be true of looking for a treasure, but it can also happen when exploring the underground to find a silver or gold mine. Finally, the anecdote of the alchemists portrays them as people who are trying to create gold themselves, rather than looking for it in the natural world. Alchemists try to compete with Nature, and to produce in a short period of time what she takes thousands of years to create. In conclusion, Constantin suggests new career paths for alchemists. Bodin implicitly has the same advice for treasures hunters, based on the notion that hard work, rather than the dangerous path of magic or Satanism, could lead them to the wealth they dream of acquiring. This progression suggests a move from a practice that attempts to use the supernatural as a means of enrichment by looking for hidden precious things, either buried treasures or silver and gold mines. The next stage is the emulation of natural processes, as in the case of alchemists. What the text suggests as the legitimate way to deal with underground matter is a new practice that would eschew any supernatural help, without trying to imitate nature.

A Practical Epistemology of the Underground

We should not understand this argument as a way for Bodin to gesture toward a modern way of engaging with the underground, as opposed to the treasure seekers' superstitious endeavours. The nature evoked here is not yet the post-bifurcation, purified realm that Whitehead and Latour identify as the modern concept of nature. This is attested not merely in the fact that Bodin does not deny the reality and workings of spirits, but also in one interesting element of the tale of the alchemists. This anecdote insists on the idea that there is indeed a vitality of matter—and of gold in particular—whose processes the alchemists were trying to emulate and accelerate. It pictures a Bennettian 'vibrant matter'.[35] Bodin, in Bennett's words, does not parse 'the world into dull matter (it, things) and vibrant life (us, beings)'. Rather, as Bennett says, Bodin encourages us not 'to ignore the vitality of matter and the lively powers of material formations'.[36] Moreover,

35 Bennett, *Vibrant Matter. A Political Ecology of Things*.
36 Ibid., p. vii.

nature (as conceived by Bodin) cannot be fully known because its laws can be circumvented by God, often through demons, as happens when a sudden violent wind destroys a building when the seekers try to excavate a treasure. The laws of Bodin's nature only hold as long as God lets them, as Jean Céard explains:

> Quand les choses arrivent 'selon l'ordre et teneur de nature', c'est que la puissance divine, faisant agir les bons démons, nous marque sa bienveillance; dès que cet ordre se trouble, si légèrement que ce soit, c'est que Dieu, faisant agir ou laissant agir les malins démons, nous manifeste sa colère et nous admoneste.[37]
>
> (When things happen 'according to the order and content of nature', it is that the divine power, making the good demons act, shows us its benevolence; as soon as this order is disturbed, even slightly, it is because God, making or letting the evil demons act, expresses his anger and admonishes us).

This highlights the impossibility of understanding Bodin's intervention as an episode in the grand narrative of the Renaissance from an age of ignorance to an era of enlightenment. The difference between the alchemists and the treasure seekers on the one hand and the honest miners on the other is not a divide between archaic endeavours and modern work. The real distinction is moral and political: for Bodin, the recourse to spirits, magic, and witchcraft was undoubtedly the main division between the two, not because he saw the treasure seekers as superstitious fools and the miners as rational agents, but because he contrasted those who used magic and witchcraft despite their illegitimacy, and those who chose a more wholesome path to success and wealth.

Indeed, within this same worldview, different types of knowledge and of engagement with nature and matter can prevail, as the stories of Agricola and Constantin make clear. These anecdotes, beyond their parallels with the previous tales of treasure seekers, play a pivotal role in the text, helping to deploy differently the arguments made in the chapter. They ostensibly conclude the part of the text devoted to the idea that the riches promised by the devil are an illusion and that his assistance does not lead to acquiring wealth. More remarkably, they segue to the next topic (namely that Satan is useless when one wants to obtain another desirable attribute, i.e. knowledge),

37 Céard, *La Nature et les prodiges*, p. 360.

so seamlessly that they can be read as an introduction to that part of the chapter—an introduction which pre-emptively modifies the focus of the developments on learning, and, at the same time, prolongs the reflection on nature, matter and things.

The few paragraphs devoted to science express a familiar lesson: it can only come from God, and never from the devil. To illustrate and bolster his view, Bodin evokes great figures such as the Latin writer Lucilius, the Biblical King Solomon, and the rabbi and philosopher Moses Maimonides (1135 or 1138–1204). The science that Bodin endorses when he turns his attention explicitly to that topic is thus high culture, classical learning, and religious wisdom. However, as the passages quoted earlier make clear, the text had already approached the idea of knowledge when treating the search for riches and treasures, although of a very different type from the high classical learning related to wisdom and religion. Through the twin figures of Georgius Agricola and Constantin, Bodin puts forward another form of learning: that is, practical, technical knowledge. Its aim was not the attainment of wisdom, but mastery over things, nature, and the acquisition of legitimate riches. The purview of those who work with their hands and not exclusively with their minds, this type of knowledge was considerably lower in prestige and honour in the Aristotelian scheme that had mostly dominated European culture since Antiquity and through the Middle Ages; in that scheme, *episteme* (or theoretical learning about unchanging things) was the most exalted, while *techne*, which aimed at making things and producing effects, was the lowest.[38] This latter form of knowledge was usually not written down, but rather passed on orally by craftsmen. However, this situation came to change, gradually but decisively, in the early modern period, as part of the series of transformations that came to be known in European history as the Scientific Revolution. Georgius Agricola, in particular, whose Latin treatise *De re metallica* 'endeavoured to ennoble an otherwise dirty enterprise', is a leading figure in this epochal transformation, and helped bring higher prestige for practical knowledge.[39] Through Agricola and Constantin, Bodin highlights in particular the knowledge that deals with matter which, like the buried treasures, is located under the Earth's surface. Bodin suggests that exploring the subterranean realm, as the practical and newly respectable science of metallurgy allows people to do, can lead to wealth that is legitimately pursued, contrary to the activity

38 On this question see Ash's analysis and references in 'Introduction: Expertise and the Early Modern State'.

39 Principe, *The Scientific Revolution. A Very Short Introduction*, p. 115.

of the alchemists who, like the treasure seekers, are trying to find morally and legally dubious shortcuts to wealth.

Bodin's text moves from buried treasures to precious metals, and from magic and witchcraft to alchemy's attempt to unnaturally speed up natural processes. The story recounted by Constantin focuses attention more closely on the question of dealing with matter and its potential to create wealth, distinguishing between legitimate and illicit ways of extracting riches from the earth. It ends up highlighting the best process in order to acquire wealth: work. Specifically, this is a type of work that does not try to emulate nature, but is rather a purely human type of work, and a distinctively human way of dealing with nature, that purposefully avoids spirits and the supernatural, without ever denying their reality. In a parallel fashion, that part of the text underlines modes of knowledge that were at that time only beginning to acquire scholarly currency—practical, non-theoretical, knowledge, as signalled by the presence of Georgius Agricola and Constantin, the geologist and the pyrotechnician.

Conclusion

Despite this interest in technical knowledge, and however enlightened his political views might be seen, Bodin, like his treasure seekers, inhabits a nature that is entirely worked through by demons, good or evil, allowed to act by God. I have shown how the Latourian concept of networks helps us to fruitfully parse the different ways in which his stories help Bodin think through the ways in which humans, things, and supernatural beings interact. Reading Bodin, and other Renaissance authors, in return, could lead us to further explore the Latourian opposition between the modern and premodern views of nature, by reminding us how deeply contested the pre-bifurcation world was itself. Jean Céard's magisterial study of the evolution of the idea of nature already cited would be of great help: it closely follows the slow change in the idea of nature that took place in the sixteenth century, while completely eschewing what he calls in another study a 'heroic view' of the history of knowledge, which tends to simply distinguish between the precursors of future enlightenment and the remnants of a benighted medieval time.[40] Céard also shows the deep crisis of the idea of nature toward the end of the century in France, which partly explains the remarkable recrudescence of demonological thinking of which Bodin's

40 Céard, 'Médecine et démonologie: Les enjeux d'un débat', p. 97.

Démonomanie is a part. Rather than being a mere representative of a simple pre-bifurcation world, the *Démonomanie* itself is a moment in a heated argument which led, according to Céard, from seeing nature as a system of signs to conceiving it as a world of things.[41] Indeed, the kinds of networks described by Bodin and his contemporaries also entail conceiving of things as something other or more than mere 'things': they are also always signs and symbols. The capacity to symbolize is another aspect of the vitality of matter in the premodern world. That dimension will be lost, as part of the transformation that will lead to radically separate nature from culture.

Works Cited

Agricola, Georgius, *De animantibus subterraneis liber* (Froben, 1599).

Ash, Eric H., 'Introduction: Expertise and the Early Modern State', in *Osiris* 25 (2010), 1–25.

Bennett, Jane, *Vibrant Matter. A Political Ecology of Things* (Durham: Duke University Press, 2010).

Bercé, Yves-Marie, *À la recherche des trésors cachés du XVIème siècle à nos jours* (Paris: Perrin, 2004).

Blair, Ann, *The Theater of Nature. Jean Bodin and Renaissance Science* (Princeton: Princeton University Press, 1997).

Bodin, Jean, *De la Démonomanie des sorciers,* ed. by Virginia Krause, Christian Martin, and Eric MacPhail (Geneva, Droz, 2016).

Braun, Christopher, *Treasure Hunting and Grave Robbery in Islamic Egypt. Textual Evidence and Social Context*, Warburg Institute, School of Advanced Study, University of London, March 2017.

Céard, Jean, *La Nature et les prodiges: L'insolite au XVIe siècle, en France* (Geneva: Droz, 1977).

———, 'Médecine et démonologie: Les enjeux d'un débat', *Diable, diables et diableries au temps de la Renaissance* (Paris: Touzot, 1988), pp. 97–112.

Dillinger, Johannes, *Magical Treasure Hunting in Europe and North America, A History* (New York: Palgrave Mcmillan, 2012).

Harman, Graham, *Prince of Networks. Bruno Latour and Metaphysics* (Melbourne: re-press, 2009).

Jacques-Chaquin, Nicole, 'La "Démonomanie des sorciers", une lecture politique de la sorcellerie', in *Jean Bodin. Nature, histoire, droit et politique*, ed. by Charles Zarka (Paris: PUF, 1996), pp. 43–70.

41 Céard, *La Nature et les prodiges*, p. 492.

Krause, Virginia, 'Listening to Witches: Bodin's Use of Confession in *De la Démonomanie des Sorciers*', in *The Reception of Bodin*, ed. by Howell A. Lloyd (Leiden: Brill, 2013), pp. 97–115.

Latour, Bruno, 'An Attempt at a "Compositionist Manifesto', *New Literary History*, (2010), 471–490.

———, *Nous n'avons jamais été modernes, Essai d'anthropologie symétrique* (Paris: La Découverte, 1991).

———, *Pandoras's Hope: Essays on the Reality of Science Studies* (Cambridge: Harvard University Press, 1999).

———, *We Have Never Been Modern*, trans. by Catherine Porter (Cambridge: Harvard University Press, 1993).

Mackenzie, Louisa, 'It's a Queer Thing: Early Modern French Ecocriticism', *FLS* 39 (2012), 15–42.

Principe, Lawrence, *The Scientific Revolution. A Very Short Introduction* (Oxford: Oxford University Press, 2011).

Sallmann, Jean-Michel, *Chercheurs de trésors et jeteuses de sort: la quête du surnaturel à Naples au seizième siècle* (Paris: Aubier, 1986).

Usher, Phillip John, *Exterranean: Extraction in the Humanist Anthropocene* (New York: Fordham University Press, 2019).

Whitehead, Alfred North, *The Concept of Nature* (Cambridge: Cambridge University Press, 2015).

About the Author

Oumelbanine Nina Zhiri (Professor of French Literature, University of California, San Diego) primarily researches questions of border crossing in the early modern period. She is the author of numerous books, including: *L'Afrique au miroir de l'Europe. Fortunes de Jean Léon l'Africain à la Renaissance* (Droz, 1991); *Les Sillages de Jean Léon l'Africain du XVI° au XX° siècle* (Wallada, 1995); *L'Extase et ses Paradoxes, Essai sur la structure narrative du 'Tiers Livre'* (Champion, 1999), as well as various articles and book chapters.

10. Reading Olivier de Serres circa 1600: Between Economy and Ecology

Tom Conley

Abstract

Formerly belonging to the literary canon of the French Renaissance, and often associated with the ideology of a return to the country—even to Maréchal Pétain's *Travail et Patrie*—Olivier de Serres's *Théâtre d'agriculture et mesnage des champs* (1600) remains a keystone in the history of agronomy. Threading the wisdom of ancient authors through his own experience, and staunchly Protestant in vision, Serres sets an agenda for the country gentleman and farmer. At once art and science, it deploys a limpid and vigorous style to argue for economy and productive management of the earth. This essay contends that today, despite its legacy, the work offers a vision and a savoury mode of writing vital to what we can make of ecology in the early modern age.

Keywords: early modern economy and ecology, Olivier de Serres, agronomy, science of gardening, return to the land, solitude

In this collection of essays, the Renaissance is located at the conceptual threshold of the Anthropocene. In French Studies, canonical authors express fears about the future of the planet that, while often set in a millenarian frame, anticipate or chime with ours.[1] In the late chapters of the *Quart Livre,* the world under the ugly regime of the well-named Messer Gaster is going to seed. In the thick of the Wars of Religion, Ronsard decries the violence that fellow subjects have done both to the Americas and the forests of his

1 For certain aspects of early modern France's relationship to the planet, see Goul, 'The Vanity of Ecology: Expenditure in Montaigne's Vision of the New World' and Usher, *Exterranean: Extraction in the Humanist Anthropocene.*

Goul, P. and P.J. Usher (eds.), *Early Modern Écologies. Beyond English Ecocriticism.* Taylor & Francis Group, 2020
DOI 10.5117/9789462985971_CH10

homeland. In a celebrated passage of 'Des coches', following Lucretius, Montaigne writes tersely and typologically, '[l]'univers tombera en paralisie; l'un membre sera perclus, l'autre en vigueur. Bien crains-je que nous aurons bien fort hasté sa declinaison et sa ruyne par nostre contagion' ('the universe will fall into paralysis; one member will be shriveled, the other vigorous. I daresay that we will have strongly hastened its decline and ruin by our contagion').[2] For Montaigne, as if they belonged to a *theatrum mundi* or world-theatre, rampant depredation and ecological ruin were signs of the end of the world he had known.

Such was the context from which Olivier de Serres (1539–1619) emerged to be read here as a thinker concerned with human reshaping of the planet. Printed on street signs in many urban centres, his name has a vaguely familiar ring. Like those a regime selects to remind distracted pedestrians and drivers of the glorious figures of its past—Monge, Thiers, Pasteur—his name figures in many urban centres.[3] A sketch of his life indicates why: born into a minor noble family whose roots extend back to the thirteenth century, he was raised at Villeneuve-de-Berg in the Vivarais and later studied at the University of Valence. In accordance with his reformed faith, he also studied in Lausanne with his brother, Jean de Serres (1540–1598), who became an eminent Protestant historian. At age eighteen, Olivier purchased the Pradel Estate, not far from his birthplace. In his *Histoire universelle* (1626) Agrippa d'Aubigné noted that in 1573, eleven years after the outbreak of the first of the Wars of Religion, Serres virtuously massacred nineteen Catholics and thus played a significant role in winning Villeneuve over to the Protestant cause. After the edict of Nantes, Serres turned Pradel into a 'theatre' of agricultural experiment, which soon became a lifelong project that included the development of sericulture, a craft for which in 1599 he made formal appeal to the king and his entourage.[4] In the following year, Protestant printer Jamet Mettayer published in folio the first edition of the compendious *Le Théâtre d'agriculture et mesnage de champs.* In 1603, Serres returned to Paris to oversee the publication of a second edition (at Abraham Saugrain) and, soon after, went back again for another (issued by Saugrain in 1605). In sum, eight editions were printed in his lifetime.

2 Montaigne, *Essais*, p. 1018 and *Essays*, p. 693.

3 The Parisian street of his name intersects with the rue de Vaugirard in the 15th arrondissement of Paris. It also is found in Lyon, in Clermont-Ferrand and Bourg-Saint-Andéol in Auvergne, in Carpentras (Provence), and twenty other centres.

4 The appeal is made in the preface to *La cueillete de la soye, par la nourriture des vers qui la font. Echantillon du Theatre d'agriculture d'Olivier de Serres seigneur du Pradel* (Paris: Chez Jamet Mettayer, 1599) [Harvard Houghton Library GEN (*FC5 Se684 599c)].

The history of *Le Théatre* indicates how it has grown out of a deeply rooted tradition of praedial literature to become a work of ecological consideration.[5] In Rouen, Geneva, and Lyon, eleven more editions of the *Théâtre* appeared between 1623 and 1675. The book then disappeared until two more editions were published in 1802 and 1804. In 1941 an anthology titled *Olivier de Serres, père de l'agriculture française* appeared with a prefatory letter by Maréchal Pétain (from Vichy, dated 18 March 1941, and addressed to the Minister of Agriculture Pierre Caziot). Thereby associated with the Vichy regime, following the war *Le Théâtre d'agriculture* was again left, as it were, fallow. Slatkine (Geneva) reprinted an edition in two volumes in 1991 before Actes Sud issued a hefty paperback (coming in at 1064 pages) in 1996; this was reprinted in 2001 (this time expanding to 1548 pages) with Pierre Lieutaghi's copious introduction. These editorial fortunes suggest that the *Théâtre* went hand-in-hand with a 'return to the land' after the Wars of Religion. Perhaps during the tumult of *jacqueries* and the two Frondes in the reigns of Louis XIII and regent Mazarin, *Le Théâtre*—in affirming its Protestant cause—was a work of war and peace. It evoked war to battle for a Reformed way of living throughout rural France; it advocated peace in mixing matters of agronomy, science, and experience while also, in a general vein, conveying wisdom drawn from country. In the early nineteenth century, it might have taken on a romantic aura and, concomitantly, been invoked to rally the nation for the ends of a populist doctrine. In the twentieth century (notably on the part of biographers during the *entre-deux-guerres*), it would mesh with the world of Charles Maurras and the Action Française. Under the Occupation, meanwhile, its aspect as a practical manual of rural life would serve the ideal of *Patrie, Famille, Travail.*

With its wealth of local knowledge (and a vocabulary no doubt greater than those of Rabelais or Zola), the *Théâtre* takes on new resonances in the twenty-first century, for it can be seen to display an ecological consciousness, potentially serving both liberal and conservative agendas.[6] The oscillation

5 Serre's work is considered revolutionary. Its mix of science and experience, what Lieutaghi calls 'the reasoned empiricism of the man from Pradel' (p. 17), set Serres apart from Columella, Pliny and Palladius (who figure in the work), and no less from Charles Estienne's *Maison rustique* of 1570, a capital study appearing in French in 1570 or, earlier and also of theological inflection, Pietro Crescenzi's *Opus ruralium commodorum,* in French as *Prouffits champestres et ruraulx, touchant le labour des champs, vignes et jardins* (1540).

6 In *Olivier de Serres: 'Science, experience, diligence' en agriculture au temps d'Henri IV,* Henri Gourdin underscores how the economy of the *Théâtre* owes to the impact of the Wars of Religion. Following a concise chronology of Villeneuve, of Serres and his family, and the fortunes of Pradel

marks Pierre Lieutaghi's informative introduction to the latest edition. On the one hand he proclaims, contrary to this volume's emphasis:

> Qu'on n'aille pas [...] chercher dans le *Théâtre* une anticipation de pensée écologique [...]. Ce n'est pas parce que l'œuvre témoigne d'un rapport à la terre aujourd'hui perçu comme 'naturel' qu'elle s'édifie dans une attention explicite à la conservation et au devenir de la nature. Quant à l'implicite, il n'est pas plus en phase directe avec le souci environnemental de l'homme du XXIe siècle.[7]

> (Let us not [...] try to see the *Théâtre* as anticipating ecological thought [...]. Just because the work bears witness to a certain relationship to the earth, perceived today as 'natural', does not mean that it pays an explicit attention to conservation or the fate of nature. As to that which remains implicit, the work in no way concerns itself with the environmental worries of twenty-first century man).

On the other hand, in the final pages, where he sees *mesnage* of the age of Serres giving way to 'management' in ours, Lieutaghi decries the industrialization of agriculture: an overpopulated Earth is nourished by cows more doped and drugged than the bikers of the Tour de France; meat now becomes 'animal-flour' of the very texture of the matter, mixed with growth hormones; steers are fed to be fattened; pigs are engineered with feet unfit to carry their bulk; machines in fields of sunflowers tear up the arable earth with consequences far more damaging than that wrought by human means; reduction of labour in fields turns soil into spongy matter designed not to hold roots but to absorb chemical fertilizer.[8] In his close reading of the *Théâtre* he declares that 'quand on regarde un peu entre les sillons de mots, on trouve une quantité de miroirs à la simplicité abrupte, où bien des visages de notre agronomie ne sauraient se voir sans honte' ('when we look closely between the furrows of words [in the book], we find a large number mirrors of an abrupt simplicity, in which many faces of today's agronomy could not look at themselves without shame').[9] And, further:

(including its addition to the list of official monuments in 1993) he lists an ample bibliography of primary and secondary materials, including seven biographical entries from 1924 to 1943.

7 Lieutaghi, 'Introduction' in Serres, *Le Théâtre d'agriculture et mesnage des champs*, ed. Lieutaghi, p. 42.

8 Ibid., pp. 48–53.

9 Ibid., p. 49.

> Cependant, toute dépassée qu'elle puisse sembler à première vue, l'œuvre transmet des connaissances pour beaucoup toujours valides, au moins dans leur principe, certaines susceptibles de nouvelle écoute. Si on considère hors de toute prévention productiviste ou scientiste, on tient un noyau d'évidences indestructibles, assez puissant pour mettre en déséquilibre bien des savoirs surajoutés, assez lourd pour peser encore dans la réflexion sur une économie rurale supposée convenir à un terroir de dimension planétaire.[10]
>
> (However, as outdated as it may seem at first glance, the work transmits knowledge that is still valid to many, at least in principle, and some of which is worthy of renewed consideration. If we set aside attempts to seek out specifically scientific and production-driven elements, then we hold here a core of irrefutable conclusions, a powerful enough core to shake up much more recent knowledge, and weighty enough to affect how we think about rural economy supposed to be suitable for a planet-size *terroir*).

In its history, then, from one standpoint the *Théâtre* resembles a hexagonal fortress defending and illustrating the riches of France for the French, while from another it is a boundless garden whose furrows are paths that lead to redemption.[11] Once honoured among literary historians of the Third Republic, today unclassifiable, but still a point of reference, we ask: what can be done with it? Survey it in a history of ecology? Read it as philosophy pre-empting the Cartesian method? Call it a cornerstone in the horticultural canon? Or, given the grace of its style and manner, set it next to *L'Astrée,* a pastoral *roman pratique*? At very least, can it figure as something other than a line in the history of agronomy?

To sort through what kind of early modern ecologies are at stake in this work, let us here follow several connected lines of enquiry. The first (1) studies the preface, a program and a guide, a *mode d'emploi* that inflects the copious work that follows. The second (2) examines the *form of the content* of the *Théâtre* in its first edition, in which the work—seen as a map, atlas and compendium—plots a diagram and a plan for both practical and ideal modes of living. A project destined for the king to behold (or hear as it is read aloud to him), the first edition relates much of its content through visual

10 Ibid., p. 53.

11 Such is the pattern of furrows that Holbein engraves in the 'Plowman', in *Les Simulachres et histoires faces de la* mort (1538), studied in my *Errant Eye: Poetry and Topography in Early Modern France*, pp. 110–113.

and textual aspects that the Protestant printer Jamet Mettayer confers upon it—notably in the title page, in the charts and illustrations at the head of each chapters, and in the form of the referential apparatus. Equally (3), the grain of the prose itself, didactic and homiletic but also textured, rife with words and usages from the *terroirs* (lands) it describes and studies, turns the manner of the *Théâtre* into its matter where (4) figuration inheres in the words themselves. Finally (5), anticipating literatures of the classical age, the *Théâtre*'s final pages establish a relation between war, retreat, and solitude that becomes the enclosing wall of the work's mental structure.[12]

The Preface

For the horticulturalist, a beginning evokes what Francis Ponge calls an *entame,* an inaugural slice or separation at the origin of a poem. It is an *ente*, a way to *enter*, a rift which puts the writing to work, whence the site from where the book will grow, ramify, and find coherence and find on the part of its owner fruitful *entretenement*.[13]

> COMME LA TERRE est la mere commune et nourrice du genre humain, & tout Homme desire de pouvoir y vivre commodément: De mesmes, il semble que la Nature ait mis en nous, une inclinaison à honorer et faire cas de l'Agriculture; pource qu'elle nous apporte liberalement abundance de tout ce dont nous avons besoin pour nostre nourriture & entretenement. D'où est venu, que, comme l'on represente soigneusement par escrit ce qu'on aime, il n'y a eu escrits ni plus anciens, ni en plus grand nombre, que de l'Agriculture; ainsi qu'on peut voir par le long denombrement des Autheurs, qui, en tous siecles & en toutes nations, ont trauaillé en ceste matiere, tres excellente & pleine d'admiration, pour l'infinie quantité des exquis & divers biens, que par elle Dieu donne à ses enfans.

12 Chandra Mukerji notes that Serres aimed to design a model of estate management, 'to integrate all farm activities within a larger rational system directed toward land improvement and profit', whose most innovative area was in techniques of the garden, in cultivating exotic and new fruits and vegetables, plotting orchards, building hotbeds, mulching, using bell jars and greenhouses (*serres*) to protect delicate varieties of garden crops, and incorporating the farm, garden and ambient countryside in an economical 'spatial system'. In *Territorial Ambitions and the Gardens of Versailles*, pp. 158–159.

13 *Entame, entamer*: the 'cut' that begins the work is its literal incipit and entry. So also the beginning of the third *lieu*, 'L'enter de la Vigne est partie de ce mesnage' ('the graft of the Vine is part of this management': p. 193), in which the grafting, *l'entement,* takes place at site of the graft, *l'enteure* (p. 194). Page numbers refer to the 1600 edition.

> (Just as the Earth is the shared mother and wet nurse of the human race, and every Man wants to be able to live there in comfort, so likewise, it seems that Nature has endowed us with the inclination to honour and regard Agriculture, for she brings to us freely an abundance of all that we need for our nourishment and care. This is why, then, because we put down carefully in writing all that we love, there have been no writings more ancient nor more numerous than those about Agriculture. This we can see from the long list of authors, who, across all centuries and all nations, have worked on this matter, quite excellent and full of admiration, for the infinite quantity of exquisite and diverse goods, that through her God gives to his children).[14]

Containing the head of a horned satyr looking upon the text that follows—its leonine mane morphing into foliage and a horn of plenty held within the square surround or plot—a historiated majuscule C signals how the words will germinate. 'Earth' is the ground where they will be nourished or suckled and grow. A minuscule portion of the *Théâtre,* the whole of the preface that follows becomes the entry and the map of a country estate and its dependencies or adjunct components. The voice of the preface speaks from a secure autobiographical place in the Vivarais, a refuge amidst the calamities of the Wars of Religion, 'ma maison aiant esté plus logis de paix que de guerre' ('my house having been more a dwelling of peace than of war'), devoted 'à mon mesnage' ('to my household'). Grafting experiment onto inherited knowledge, he has been inspired *to write*:

> Ie dirai donc librement, qu'aiant souvent & soigneusement leu les livres d'Agriculture, tant anciens que modernes, & par experience observé quelques choses qui ne l'ont encores esté, que je sache, il m'a semblé estre de mon devoir, de les communiquer au public, pour contribuer selon moi, au vivre des hommes. C'est ce qui m'a fait escrire.

> (I will say freely, then, that having often and carefully read books about Agriculture, both ancient and modern, and through experience having observed many things that have not yet been communicated, as far as

14 *Le Théâtre d'agriculture et mesnage des champs,* 3 [fol. a.iv r]. Reference will be made both to this edition and Pierre Lieutaghi's recent edition, to which is added, *dans lequel est représenté tout ce qui est requis et nécessaire pour bien dresser, gouverner, enricher et embellir la maison rustique* (in which is represented all that is required and necessary in order to properly build, govern, enrich and embellish the rustic house: p. 59). The text here has been slightly modernized with additional accent marks. The original features historiated initials and head- and tailpieces in a style imitating the design of an earlier date.

I know, it seemed to me that it was my duty to communicate them to the public—that is my contribution to the life of men. This is what has driven me to write).[15]

The book spells out its mission by way of analogy: *just as, so then, whence* and *thus*… the Earth, our Mother Nature and implied daughter of God, nourishes us through what we produce with her by dint of cultivating soil and writing. Like assiduously managed fields, an elegantly furrowed page is a sight to behold. The comparison suggests that the book, a product of earlier science and reason in view of more recent 'experience and practice',[16] becomes a field that the reader tills with the mind's eyes and hands. The information that pioneers of agriculture, *ceste belle science* (this lovely science)[17] have passed onto succeeding generations, inspires a totalizing, even encyclopedic knowledge that the 'theatre' organizes and stages so it may be seen both in its totality and in a specific situation. France and its soil in the aftermath of the Wars of Religion become the ground where science and experience become the matter of management.

Serres's project or design (*ce mien dessein*), is to show as best he can, *brievement & clairement*, how to cultivate '*la Terre* & ce pour commodément vivre avec sa famille' (the Earth—in order to live comfortably with his family).[18] The land is a living body that industrious landowners can cause to 'give birth to its fruits' (from 'its entrails') that serve all living needs. The proprietor of a carefully chosen land must follow the precepts of two diagrams or memory-images: the first correlating *Art*, the fruit of *Experience*, with good judgment and use of *Reason*; the second, meanwhile, relies on putting *diligence* to good use to show that the knowledge (*science*) of agriculture is much like experience (experiment with and transmission of local or 'time-tested' knowledge). As if pre-empting Balzac's famous symbolic triangle in the early pages of *La Peau de chagrin*, he writes in bold letters, 'Que pour faire un bon Mesnage, est necessaire de joindre ensemble le SÇAVOIR, le VOULOIR, le POVVOIR. En ceste liaison gist l'Usage de nostre agriculture' ('For a good household/estate/husbandry, one must combine Knowledge, Will, Ability. In this combination lies the Way of Agriculture').[19]

15 Serres, *Le Théâtre d'agriculture et mesnage des champs*, fol. a.iv v; ed. Lieutaghi, p. 60. Subsequent references will be made to the first edition (1600), followed by Lieutaghi (2001).

16 Ibid.

17 Ibid., fol. a.iv v and p. 61.

18 Ibid., fol. e r and p. 61.

19 Ibid., fol. e.iii r and p. 65. And Balzac might have been citing *Le Théâtre*: in the opening pages, the magus in the antiquity shop utters to *l'inconnu* (the young man, soon after named as

'[L]e plan general de tout ce grand Discours' ('The general plan of this whole Work')[20] is divided not into chapters but into eight *Lieux* or 'places'. The design becomes part and parcel of the dwelling, the author's estate, but also the estate understood as a house of being: a dwelling, perhaps, in the sense of the concept of *Dasein.* The folio format of the first edition is crafted to resemble a dwelling unto itself, to be a proprietor's atlas, a practical Bible of Reformed inspiration, a self-enclosing narrative, and a composite work whose science of experience is both legible and visible. Hence the importance of the design of its images and textual disposition: principally the frontispiece, the title pages of each *Lieu,* and the Ramist diagrams that follow them; additionally, in the sixth *lieu,* the illustrations depicting two hypothetical gardens. The disposition of text, proverbs, and *manchettes* turn the material into memory-places.

The Frontispiece, Title-Pages, and Diagrams

Seen in slight bird's eye view, in a finely drawn copperplate engraving, a great triumphal arch contains the title, the name of the author, and the date of publication (Figure 2).[21] Two columns of alternating blocks of dark and white stone, recalling the style of Henri IV and resembling the columnar designs of Philibert Delorme, support the platform on the top of the arch; eight parterres and eight gardens, each of a different design, stress the rectilinear or gridded aspect of a space at whose centre is a statue of Henry, seated Christ-like on a throne, holding a sceptre in each hand.[22] Three allegorical figures—Justice, Peace, and a female figure who would seem to

Raphael de Valentin), who is on the verge of suicide: 'Deux verbes expriment toutes les formes qui prennent ces deux causes de mort: VOULOIR et POUVOIR. Entre ces deux termes de l'action humaine, il est une autre formule dont s'emparent les sages, et je lui dois le bonheur et ma longévité. *Vouloir* nous brûle et *Pouvoir* nous détruit; mais SAVOIR laisse notre faible organisation dans un perpétuel état de calme' ('Two verbs express all the shapes and forms that entail these two causes of death: WILL and POWER. Between these two terms of human action, there is another formula that the wise make use of, to which I owe my happiness and longevity. *Will* burns us and *Power* destroys us; but WISDOM leaves our weak organization in a perpetual state of calm'): *La Peau de chagrin,* ed. Satiat, pp. 88–89.

20 Serres, *Théâtre,* fol. e.iii r and p. 65.

21 Mortimer (*Catalogue of Books and Manuscripts,* p. 606) notes that the upper portion of the design illustrates the virtues he has brought to the nation and that, as the signature in the lower corner indicates, Karel van Mallery (1571–1645), well known Flemish engraver working for Jamet Mettayer, is the author of the illustration.

22 On Delorme's architectural economy in his work on columns see Zerner, *L'Art de la Renaissance en France: L'Invention du classicisme,* pp. 373–389.

Figure 2: Frontispiece from Olivier de Serres, *Le Théâtre d'agriculture et mesnage des champs* (Paris: Jamet Mettayer, 1600). Image from Houghton Library, Harvard University (Cambridge, MA).

be Agriculture—are in his midst. At each corner, an urn filled with sprouting flowers draws attention to festoons of legumes on either side of the fronton, on which is written, above the key to the vault below, *SECURITAS PVBLICA*. The woman in the foreground who digs her spade into the soil, personifying the management of the garden (the garden that is France itself) is also the allegory of Public Safety. In the two adjacent spandrels are bundles of the farmer's and gardener's instruments—basket, scythe, garden templates, clippers, sickle, spade, fasces—and in the two compartments above the plinths are two views of country houses, whose smoking chimneys signal warmth and order. The arch itself is set in a landscape, whose background includes mountains with houses on their slopes; in the middle ground, fields give way to trees: on the left there is an olive tree or *Olivier*, whose limbs are the icon of peace, the branch that the allegorical figure above embraces with her left arm. On each side, a corner of a well-kept wheat field suggests that growth and production extend beyond and behind the arch. The parterres on the top anticipate the garden design that become synonymous with the king's innovations at Saint-Germain-en-Laye and at Fontainebleau, while the three allegorical figures respond to the words of the dedicatory letter addressed to the king. The letter praises him for the peace he has brought to the nation that now lives in 'seurté publique sous son figuier, cultivant sa terre, comme à vos pieds, à l'abri de Vostre Majesté, qui a à ses costés la Iustice et la Paix' ('public safety under his fig tree, cultivating his land, as if at your feet, in the shade of Your Majesty, who has by his sides Justice and Peace') and for having delivered his people 'from the fury and fear [*fureur & fraieur*] of the cruel wars'.[23]

The frontispiece underscores rectitude of perspective, geometry and linear order; at the same time, its allegorical design anticipates the textual and visual plan subtending the eight *Lieux*. Each features a title page on which an oblong woodcut depicts the settings of the given places, respectively: the well and courtyard of the farm (1), a landscape whose fields are being tilled and sown with seeds (2), a vineyard on a hillside adjacent to a grange holding a press (3), a field where women gather wheat and milk a cow (4), a chicken-yard and a man and woman ordering beehives (5), a garden of parterres and grove trees where labourers work (6), the edge of a forest where men are felling and limning trees (7), and a landscape where men and women are hunting (8). The cuts summarize much of the content of the copious textual matter over which they stand, each picture a visual

23 Serres, *Théâtre*, fol. ã.ii r and p. 55.

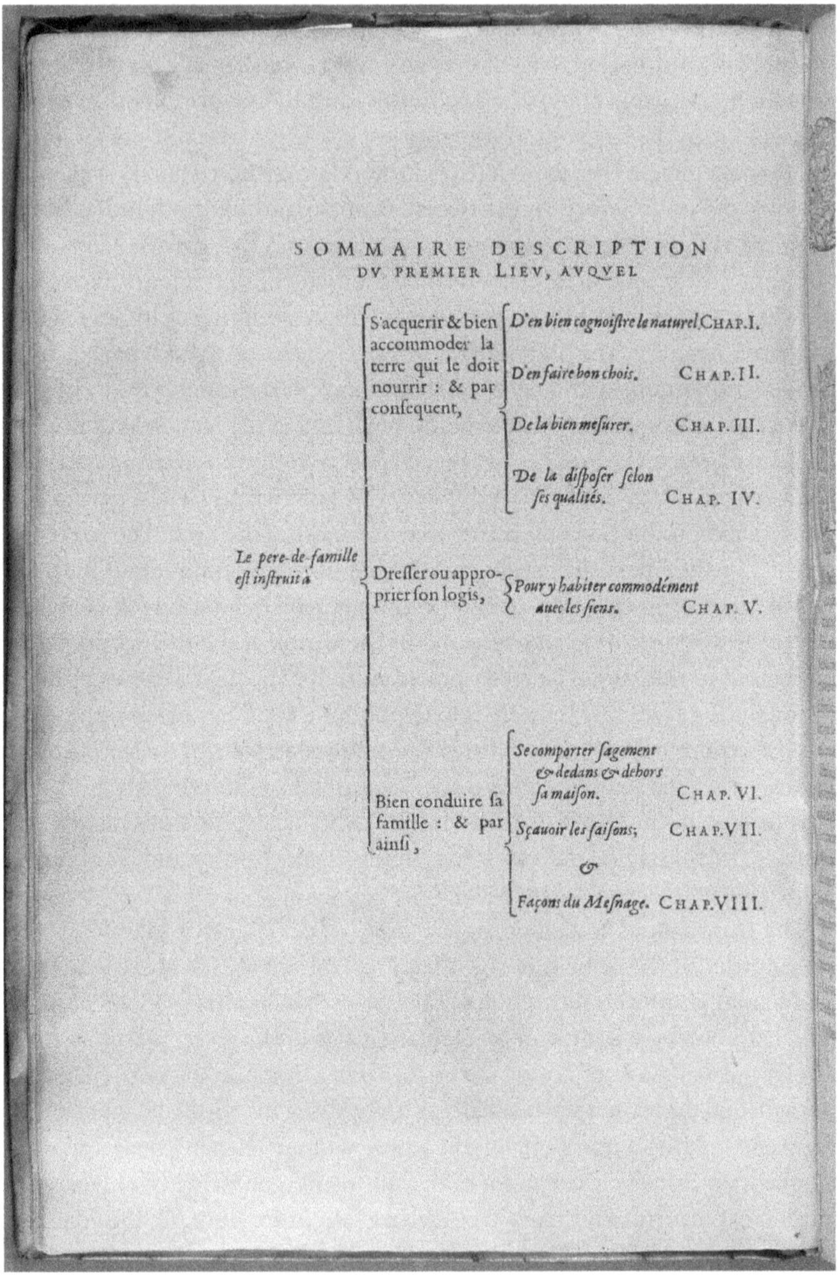
SOMMAIRE DESCRIPTION
DV PREMIER LIEV, AVQVEL

Le pere-de-famille est instruit à

S'acquerir & bien accommoder la terre qui le doit nourrir : & par consequent,
D'en bien cognoistre le naturel. CHAP. I.
D'en faire bon chois. CHAP. II.
De la bien mesurer. CHAP. III.
De la disposer selon ses qualités. CHAP. IV.

Dresser ou approprier son logis,
Pour y habiter commodément auec les siens. CHAP. V.

Bien conduire sa famille : & par ainsi,
Se comporter sagement & dedans & dehors sa maison. CHAP. VI.
Sçauoir les saisons; CHAP. VII.
&
Façons du Mesnage. CHAP. VIII.

Figure 3: Diagram of the structure of the 'Premier Lieu' in Olivier de Serres, *Le Théâtre d'agriculture et mesnage des champs*, p. 15.

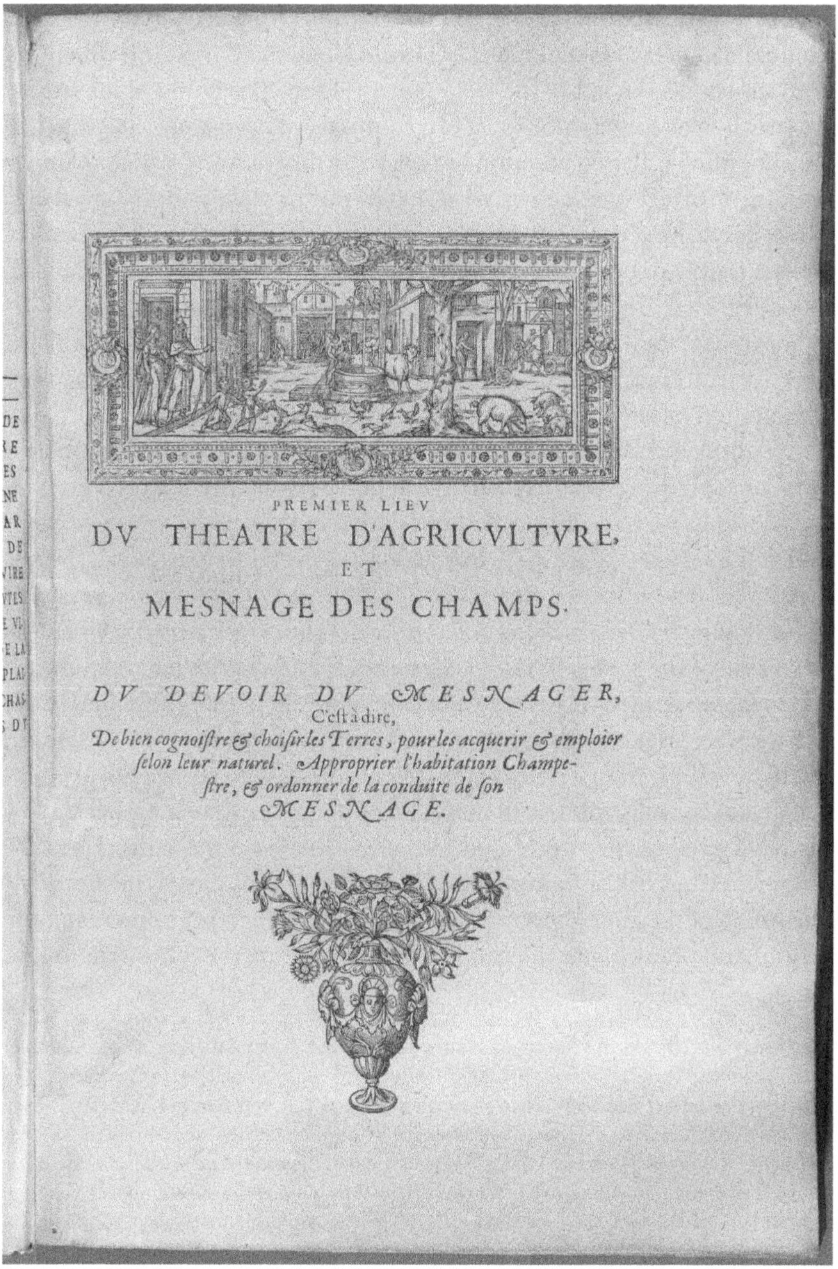

PREMIER LIEV

DV THEATRE D'AGRICVLTVRE,

ET

MESNAGE DES CHAMPS.

DV DEVOIR DV MESNAGER,

C'eſt à dire,

De bien cognoiſtre & choiſir les Terres, pour les acquerir & emploier ſelon leur naturel. Approprier l'habitation Champeſtre, & ordonner de la conduite de ſon MESNAGE.

Figure 4: Title page of the 'Premier Lieu', from Olivier de Serres, *Le Théâtre d'agriculture et mesnage des champs*, p. 16.

device, a seedbed for what the text develops in extensive detail.[24] Almost emblematic, the woodcut images belong to a mnemonic system or memory mechanism. To the left of the first 'place' (Figure 3) is found in majuscule the sculpted or epigraphic aspect of a summary of the chapter and much of the entire book. The connection between the diagram and woodcut image signals how the book is to be read. Each unit or place is designated as a parcel containing careful arrangements of practical matter, like a book of recipes that is to be consulted and read in accord with its own order and economy.[25]

The images mesh with the order of the work. Set in scenographic perspective, illustrating the contents of what will follow in the exposition, they sum up and present variation on the contents of the prose. The oblong frame of the woodcuts emphasizes landscapes—a wide-angle view, as it were, or something akin to a theatrical perspective where, in the *Premier lieu* (first place), the courtyard about which buildings are arranged has a well at the centre (Figure 4).[26] A woman pours water from a bucket she has hoisted with windlass while, in the foreground, two children approach a dog galloping by five chickens and four piglets grazing in the corner on the other side of the scene. In the background four *métayers* (sharecroppers) and *fermiers* (farmers) attend to their labours while, in the foreground to the left, the spectator gets a first glimpse of the character, the *père-de-famille*, the benevolent, industrious and diligent father who shows his prudent and loyal kin and staff how to manage the country estate. Emphasized by the wide and ornately decorated frame, the enclosed space describes the inner workings of the home, clearly a pictorial rendition of the *mesnage*. The home of the first chapter is the 'soul' of the 'theatre', a space equally mental and physical. As the chapter makes clear on the folios that follow,

24 The second edition, 'reveuë et augmentee par l'Auteur' (Paris: Saugrain, 1603), in smaller format, economizes by removing this page and the blank folios juxtaposed to the title pages of the seven *lieux* that follow. The sense of the blank page as fallow plot or ground before which the image and the matter to follow emerges or grows is missing. In the 1603 edition the smaller point-size of the typography and the crammed aspect of the printed page underscore a more practical and economical design than we see in the austere elegance of the original folio edition.

25 The book was valued, observes Mukerji, for its practical information. A 'dry and systematic bit of writing, designed to be read for reference rather than pleasure', the *Théâtre* 'was a practical bible for the literate, newly enriched gentleman of France who sought to find comfort and profit in the countryside' (*Territorial Ambitions and the Gardens of Versailles*, pp. 158–159). It might be added that the manner or style belongs to a 'literary enterprise' (ibid., p. 161), which Mukerji associates with authors aiming their works at 'upwardly mobile readers' (ibid., p. 162).

26 The scenes are briefly taken up in Antoinette Lavondès, 'Les éditions du "Théâtre d'agriculture"', *Bulletin de la société de l'histoire du protestantisme français*.

the *père-de-famille* must have the mental wherewithal to manage inner and outer spaces, and a reasonable portion of land befitting his situation.

The presentation of each chapter plays on a tension of analogy and representation. Printed on the verso of the title page of each *lieu* is a Ramist diagram, a 'sommaire description' that establishes the hierarchies of the subject matter. Unlike the comparative process developed in the text of the first section, the summary diagram to the left plots the way the property owner, again the *père-de-famille*, must organize and channel his energies.[27] The diagram is the 'science' where the text becomes its field of experience, where comparison, reminiscent of the art and experience of comparison advocated by Montaigne, is tantamount to mediation or 'management'. What is easily 'represented' or assimilated in a text-map whose origin and end is the family father brings forward an implicit textual geography. Knowledge comes with the recognition of topography both of the national space for which the book is written and, by analogy, the book itself. The work begins as if its own syntax were following the order of a diagram in response to what the eye notes on the opposite page.

> LE FONDEMENT de l'Agriculture est la cognoissance du naturel des terroirs que nous voulons cultiver, soit que les possedions de nos ancestres, soit que les aions acquis; afin que par ceste adresse, puissions manier la Terre auec artifice requis; & emploians à propos & argent & peine, recueillons le fruict du bon mesnage, que tant nous souhaitons: c'est à dire, contentement auec moderé projet & honeste plaisir.
>
> (The foundation of Agriculture is the knowledge of the natural qualities of the lands that we want to cultivate, whether we inherited them from our ancestors, or whether we acquired them; in order that through this knowledge, we may handle the Earth with requisite artifice; putting into it both money and great effort, we can reap the fruits of good care, as much as we like: that is to say, contentment with moderate intention and honest pleasure).[28]

27 Serres, *Le Théâtre d'agriculture et mesnage des champs*, fol. 2v and p. 71.

28 Ibid. Where Serres distinguishes between lands inherited and acquired faint echoes of the beginning of Machiavelli's *Prince* can be heard. The third chapter that deals with mixed principalities studies how a prince can manage them. From the beginning the ideal father of the work would be the pastoral analogue to *The Prince*. Machiavelli and Serres share a genealogy through Charles Estienne, author of *La Maison rustique* (1564), who printed Guillaume Cappel's translation of Machiavelli: *Le Prince de Nicolas Machiavelle secretaire et citoien de Florence*. The incipit to 'De l'office du pere-de-famille enuers ses domestiques, & voisins' ('Of the role of the

In the following sentences, comparative differences demand attention to specific places and local knowledge. 'Par là doncques nous commencerons nostre Mesnage, & dirons qu'on Remarque plusieurs & diuerses sortes de terres, discordantes entr'elles par diuerses qualités; lesquelles difficilement peut-on toutes bien representer'.[29] The linear design bends and turns when analogy intervenes, where it distinguishes soils of clay from those of sand in order to set into uncanny opposition fertility [*fertilité*] and sterility [*fterilité*], whose contrary attributes are of almost identical graphic character.

> Car comme le sel assaisonne les viandes, ainsi l'argille & le sablon estants distribués és terroirs par juste proportion, ou par Nature ou par Artifice, les rendent faciles à labourer, à retenir & rejetter conuenablement l'humidité; & par ce moien, domptés, apriuoisés, engraissés, rapportent gaiment toutes sortes de fruicts. Comme au contraire, importunement surmontés par l'une ou l'autre de ces deux differentes qualités, ne peuuent estre d'aucune valeur: se convertissans en terres trop pesantes, ou trop legeres; trop dures, ou trop molles; trop fortes, ou trop faibles; trop humides, ou trop seches; bourbeuses, croieuses, glaireuses, difficiles à manier en tout temps, craignans l'humidité en Hyuer, & la secheresse en Esté; & par consequent presques infertiles.

> (For just as salt seasons meats, so also clay and sand when distributed across lands in just proportion, either by Nature or by Artifice, make them easy to work, to hold, and to adequately divert humidity; and through this method, harnessed, tamed, fertilized, they joyously produce all sorts of fruits. Just as on the other hand, inopportunely surmounted by one or the other of these different qualities, cannot be of any value: transforming into earth too heavy, or too light; too hard, or too soft; too strong, or too weak; too humid, or too dry; muddy, chalky, viscous, difficult to manage in

paterfamilias in regards to his servants and neighbours') thus begins: 'Ces choses seroient vaines sans bon gouuernement, ne pouuant en ce monde rien subsister sans police. [...] En cela, imitant le General-d'armée, qui emploie aux fortifications, des pionniers, n'aiant, comme beufs, autre valeur qu'en la force, sans esprit ni entendement. Sur ce sujet dit le Poëte, *Que son vers chante l'heur du bien-aisé rustique,|Dont l'honneste maison semble une Republique*' ('These things would be vain without a good government, as nothing in this world can persist without rule. [...] In this, imitating the army general, who employs in the construction of fortification foot soldiers who, like oxen, have no other value than strength, without spirit or reason. On this subject the Poet says: May his verse sing the time of the happy rustic | Whose honest house resembles a Republic': p. 25).

29 Ibid.

Figure 5: Woodcut illustration at the head of the 'Sixième Lieu' in Olivier de Serres, *Le Théâtre d'agriculture et mesnage des champs*, p. 497.

all weather, fearing the humidity in Winter, and the drought in Summer; and as a result almost infertile).[30]

In accord with the attractions and repulsions of an Aristotelian diagram of the four elements, a variety of attributes fan out from the opposition set in the matrix of comparison.

30 Ibid. The opposition becomes a structuring agent throughout much of the *Théâtre*. A propos of the variety of oak trees in France, the observation here (mixed with a hint of erotic fantasy) is ordered according to the resemblance and difference. 'Plusieurs sortes de glands, se remarquent és Chesnes en general, lesquelles Pline met en nombre de quatorze, comptant pour vne la faine, produite par le Hestre ou Fousteau, dit en Latin, *Fagus*. Telle difference de glands, est remarquable en toutes qualities: de grãdeur, couleur, figure, pesanteur: l'abondance duquel fruict, agreste, prouient plus grande des Arbres femelles, que des masles, à telle occasion, leur sexe aiãt esté distingué par les Anciens, comme aussi de tous autres Arbres portant fruict, dont les steriles ou de peu de rapport, ont esté estimé masles, & les fertiles, femelles' ('Many sorts of acorns, generally found in oak trees, of which Pliny identifies fourteen, including the beech nut, produced by the beech tree, in Latin, *fagus*. Such a variety of acorns is remarkable in all qualities: size, colour, figure, weight: the abundance of such a fruit, fertile without human labour, comes mainly from female trees, rather than male, in such an occasion, their sex having been distinguished by the Ancients, as for any other tree bearing fruits, that the barren ones or those of little yield, have been esteemed to be male, and the fertile ones, female': p. 795 and p. 1172).

The Grain of the Prose: The Garden and its Architecture (*Sixiesme lieu*)

The sixth chapter is a point of reference in the history of garden design prior to André Le Nôtre. The diagram facing the title page indicates the arrangement and management of four types of garden (vegetable, flower, medicinal, and fruit) while, on the opposite folio, the woodcut signals how the garden and chateau are correlated (Figure 5). Set in an ornate frame, the cut portrays men and women at work along the *allées* of a plot situated between a set of buildings and an adjacent grove of trees. The ornate surround is comprised of four blocks whose corners are cut along a diagonal line. The gap between each of them indicates that they are moveable and modular—the same borders enclose each of the illustrations on the title pages of the eight *lieux*. Along the diagonal of each of the four corners emerge blank lines (of a width of one millimetre) that become four sightlines opening onto the perspective of the boxes and parterres under construction. Inside the image, the perspective leads to a trellis that serves as an entry onto a landscape where, in the far distance, a church stands atop a hillside. The visual trajectory moves through various stages of the labours before reaching virtual salvation at a vanishing point. A dog in the foreground stands on its hind legs, its head arched back to glimpse at two workers who are planting a tree destined to align with those along the border of the grove in the middle ground, all of which lead to the trellis and church in the back.[31] Coequal to the two workers on the left, a sturdy farmer, armed with a shovel, turns over the soil for what would be the fourth parterre in the middle of three *allées*. Replete with borders and designed with a lozenge and a circle at each axis, the first two parterres in the foreground are complete; the two further back are yet to be drawn. The parterres to the left (the first divided into two rectangles behind a pot of flowers) also remain fallow. Along the central sightline that moves from a cartouche in the lower border to the trellis, a kneeling woman tends to the parterre in front of her. Seen in its totality, the image presents four elements—a managed grove of trees, a garden and it accoutrements, a country house, and the landscape where the owner's tilled fields and pastures are in immediate vicinity to each other. The central *allée,* which moves directly through the space and to the

31 The posture and dress is reminiscent of Etienne Delaune's scenes of the months and seasons that belong to the manner of Fontainebleau of the 1560s. (See Delaune, *Douze mois*).

vanishing point, and toward which the workers are aimed, pulls together the components of the estate.[32]

The opening sentences signal how the parterres will figure in the garden's overall design and management. Before God's vast and rich creations, we find ourselves in need of order. The Creator,

> donnant à l'homme tant de sortes de viandes, différentes en matiere, figure, capacité, couleur, saveur, proprieté, qu'impossible est de les pouvoir toutes discerner ni comprendre. Et comment telles largesses de Dieu pourroit l'homme representer naifvement, veu qu'il n'est encores parvenu à leur entiere cognoissance, se descouvrans tous les jours de nouvelles plantes, non seulement estrangeres, ains mesme croissans parmi nous?
>
> (giving to man all kinds of foodstuffs, different in substance, shape, size, colour, taste, and properties, so that it is impossible to discern and understand them all. And how could Man in true clarity represent God's great generosity, seeing that Man does not yet know them fully—every day new plants are discovered, not only foreign ones, but also ones that grow close by?)[33]

We must identify, name, arrange, and order an ever-growing number of species, the format of the *Théâtre* suggests, in accord with spatial reason. The parterres and compartments of the garden become a decorous dictionary similar to the *Théâtre* itself. Referring to Fontainebleau, Saint-Germain-en-Laye, the Tuileries, Monceau 'and so on', Serres remarks how what is seen will be read and variously deciphered: 'Ce ne pourroit voirement estre sans merveille, que la contemplation des herbes *parlans par lettres, devises, chiffres, armoiries, cadrans*: les gestes des hommes et bestes: la disposition des edifices, navires, bateaux et autres choses contrefaictes en herbes & arbustes' ('It surely is not without marvel that we contemplate those plants that speak *via letters, devices, numbers, heraldic signs, and sundials*; that we observe the feats of men and beasts; that we look on at the ordering of buildings, ships, boats, and other things made from plants and bushes'). He adds that with such 'merveilleuse industrie & patience' ('marvellous

32 The *allée*, notes Mukerji (*Territorial Ambitions and the Gardens of Versailles*, p. 159) was designed to 'link the house to other parts of the property. This system organized the two into a common, rational (often bilaterally symmetrical) order' that would define many French gardens in the age of Louis XIV.

33 *Théâtre*, fol. R.rr.ii r and p. 781, emphasis added.

labour and patience') we need not go to Italy or elsewhere to admire 'les belles ordonnances des jardinages, puis que nostre France emporte le prix sur toutes Nations' ('the fine constitution of gardens, since our France wins out over all Nations').[34] The compartments must be disposed to maximize light, and the trees that border them never too high; if not, statues and geometrically sculpted stones can be set in their place, even bushes be tailored to resemble them. Above all, he adds, what Claude Mollet has done in the Tuileries and at Fontainebleau stands as a model of *portraiture*. The earth of the parterres must be fertilized and spaded to allow plants whose differing types and patterns of roots can take hold and provide nourishment. The different flowers, grasses and herbs are to the parterres what line and colour are to paintings. They must be carefully cut and pruned, to which attest the patterns of the twelve illustrated compartments.[35]

To describe the effect the parterres have on persons whose eyes gaze upon them, and in order to meld science, practical endeavour, and aesthetics, Serres brings together the medicinal and botanical garden. The conceit results in pleasure and profit, which are taken to be the two axes of an ellipse:

> Ce sont les ornemens du jardin de plaisir, destinés au contentement de la veue. Recréent aussi l'esprit, les précieuses et douces senteurs, procédantes d'une infinité d'herbes et de fleurs qu'on y esleve d'entendement s'employe de grande affection, pour soulagement en ses sérieuses affaires. Et comme la bonne musique ne saoule l'aureille de ceux qui l'aiment, ains, cessant, la laisse affamée: aussi le plaisir qu'on prend à voir et à odorer les herbes & les fleurs de belle representation et de bonne senteur, n'est jamais parfaict. Dont avient, que c'est tous-jours à recommencer, que le jardin à fleurs, où à toute heure l'on treuve de la besongne, soit ou pour y ad-jouster de nouvelles plantes, soit ou à agencer et entretenir les vieilles. De mesme en est-il du jardin médicinal, où le plaisir n'est petit, d'y voir infinité de plantes de diverses sortes & proprietés, qu'on esleve et entretient avec continuelle solicitude. Mais comme le bouquetier a pour premier but, le plaisir; le médicinal vise principalement au profit et à nécessité, pour le soulagement de nos infirmités. Neantmoins, ont-ils de commun ces deux jardins-ci, le plaisir et le profit.
>
> (They are the ornaments of the pleasure garden, destined to the contentment of sight. The mind finds pleasure in the precious and sweet aromas,

34 Serres, *Théâtre*, pp. 579 and 895.

35 Lieutaghi Serres, *Théâtre*, pp. 586–605 and plates 1–12 between 904–905.

> coming from an infinity of herbs and flowers that our senses are elevated, and employed with great affection for alleviating serious matters. And just as good music never tires the ears of those who truly enjoy it, similarly, when it stops, it leaves them famished: so too the pleasure we take in seeing and smelling herbs and plants beautifully displayed and redolent of good aroma is never perfect. So it goes that it is always to be started over: to the flower garden, in which we can labour at all times of day, can be added new plants or arranged so as to maintain the older ones. The same goes for the medicinal garden, the pleasure of which is not to be discounted, in the sight of infinite numbers of plants of various sorts and virtues that are maintained with continued care. But as the principal goal of the flower garden is pleasure, the medicinal garden is designed for need, profit and, no less for the alleviation of our infirmities. Nonetheless these two gardens share the virtues of both pleasure and profit).[36]

The mention of the medicinal garden leads to an uncommon visual fantasy capping the sixth *lieu.* Whereas the good family father can afford only a plot for curative plants, service to the king 'me faict un peu sortir des limites de mon intention' ('leads me to step a little outside of my intentions') by crafting two monuments, each of which is illustrated in perspectival and ichnographic views.[37] For the first, he will fashion from enriched earth, soil, and stone *une montaignette,* a little mountain in the shape of a square, whose sloping elevation reproduces the climates of the earth and whose sides, facing the four cardinal directions, will make use of sun and shade while protecting plants from the colder winds. At its 'peak', a fountain will water the plants on the slope of the five terraces, while stairwells in the middle of the sides facing the north and west and at the four corners will, '*comme un théâtre,* dont les degrés s'excédans les uns les autres, rendent le jardin de très-belle representation' ('*like a theatre*, whose steps stick out beyond each other, turn the garden into a very handsome display').[38] Like a theatre of different stages, or a defensive structure offering the assuring that peace now replaces war, and like the very architecture of *Le Théâtre* itself, the square-shaped medicinal monument will resemble a ziggurat (Figure 6).

The second model is of circular design. An upward-spiralling terrace, punctuated by little gardens resembling the ratchets of a windlass, leads to a platform at the top, at the centre of which is a fountain designed to

36 Serres, *Théâtre*, p. 905.

37 Serres, *Théâtre*, pp. 599 and 908, emphasis added.

38 Ibid., pp. 600 and 909.

water the plantings in the angled boxes on the levels below. The idea finds inspiration in both biblical myth and the author's own locale. The spiral terrace or walkway is 'semblable à celui qu'on remarque peint, en la tour de Babel: et qu'encore l'on void, à celle du far d'Alexandrie d'Egypte, basti au port dans la mer. Entre les antiquités de Nismes, y a une mazure qu'on appellee, *Tourré-magne,* qui semble avoir esté bastie à la maniere susdite' ('similar to that which can be seen, painted, in the Tower of Babel, which still can be seen, and to the Lighthouse of Alexandria, in Egypt, constructed in the harbour. Between the antiquities of Nîmes there is an old building they call *Towermagne,* which appears to have been built as noted above').[39]

In the larger scheme of the *Théâtre*, the two plans are *tours de force*. Where the circular garden strikes the eye as a mechanical form, an instrument and a tower, the square counterpart could be the model of a mausoleum, an arsenal, or a defensive pillbox. Because both structures are furnished with a fountain at the centre of a spacious platform at their upper level, the common space recalls the platform at the top of the triumphal arch at the frontispiece where the parterres and their compartments announce the portraits of the gardens in the thirteenth chapter. Analogy would posit that the fountain at the top of the king's ideal medicinal garden is replaced by the portrait of the king himself, a font of life and nourishment, adjacent to the three allegorical women who are counterparts to the workers in the illustration at the head of the sixth section.

In three key areas, figure and analogy work with and against one another. First and foremost is the relation of the Protestant author to his Protestant printer: the former, the *pater familias* of his work, employs good men to execute the tasks that keep the house in order. Such is Jamet Mettayer, inscribed on the threshold of the triumphal entry of the title page, set below 'Olivier de Serres, Seigneur de Pradel'. As 'Ordinary printer to the king', Mettayer is the *mettaier* who takes charge of the fields that his lord and author has allotted him.[40] The work presents an almost egalitarian distribution of wealth: 'I'estime la condition raisonnable, si en la cueillette, les gerbes ou les grains sont partagés par moitié entre le seigneur & le metaier' ('I esteem

39 Ibid. Pierre Lieutaghi in his introduction wonders if, in its relation to the round counterpart that resembles a Tower of Babel, this ideal classificatory model is not unconsciously related to the fact that, as Serres avows, the infinite matter that belongs to the garden cannot be pigeonholed or accounted for. Given its impossible taxonomy the garden is affiliated with an architecture of inaccessibility. The maps and figures of the medicinal garden belong to an 'exemplary fantasm' (p. 46).

40 'Metayer: a farmor, or husbandman; properly, one that takes grounds to the halves; binds himself, by contract, to answer unto him, of whom he holds them, halfe, or a great part, of the profits thereof' (Cotgrave, *A Dictionarie of the French and English Tongues*).

the condition reasonable and fair if, in the harvest, the sheaves or the seeds are equally shared by the lord and the farmer').[41] Further, the *mettaier* who would be farmer and printer is an '[h]omme de bien, loial, de parole, & de bon compte: sain: aagé de vingt-cinq à soixante ans: marié auec vne sage & bonne mesnagere: industrieux: laborieux: diligent, espargnant; sobre, non amateur de bonne chere; non yvrongne; ne babillard: ne plaideur: ne villotier; n'aiant aucune bien terrier, ou au Soleil; ains des moiens à la bource' ('a good man, loyal, who keeps his word and is accountable: healthy, between twenty-five and sixty years, married, with a wise and good manager, industrious, hard-working, diligent, economical; sober, neither given to revelry, inebriation, nor chatter, never complaining, nor running about nor a gossip; having no earthy belonging, nor under the sun; nor means in his purse').[42]

Figs and Figures

The relation between figs and figures is especially evident where the author compares a variety of trees. Can it be by chance the discourse on the fig tree is adjacent to that of the olive tree? Yes and no: in the broader poetic tradition, the fig is the icon of figuration. An emblem of the coincidence of word and thing, it assures sensuous procreation. The *figue* anticipates its *figure*, thus promising variety and productivity. The tradition of the blazon of a precious stone, a bodily part, or a fruit for which Marot, Scève, and Rémy Belleau had been past masters, otherwise foreign to the work of a practical treatise, is present, all the more because the fig invites contemplation:

> La bonté de la Figue n'est mise en dispute, chacun tenant ce fruit-là estre des plus exquis, lequel & le Raisin, par jugement vniuersel, sont estimés la couronne de tous autres. De faict, à l'arrivee des Figues & des Raisins, marchans presque de compaignie, on void disparoistre la plus grande partie des autres Fruits, comme cedans à leur délicatesse. Aucuns ajoustent à ce roole la Peche, mais il se trompent: car bien qu'elle soit bonne, s'entend des bien choisies, si est-ce que meilleures sont plusieurs Poires, Pommes & Prunes, partant indigne d'estre couchee en si haut degré. *Reuenant à la Figue, dirai qu'admirable est la contemplation de tel Fruit,* pour l'abondante variété de ses espèces & leurs diuerses qualités, duquel, ainsi que des Poires, les Anciens ont fait très-grand estat.

41 Serres, *Théâtre*, pp. 59–60 and 154.

42 Ibid., pp. 61–62 and 154.

(The goodness of the Fig is not disputed here, everyone holding this fruit to be of the most exquisite, of which, together with the Grape, and by universal judgment, they are esteemed to be the crown of all others. Indeed, upon the arrival of Figs and Grapes, marching almost in formation, disappear the greater part of all other Fruits disappears, as if they yielded to its delicacy. Some add to this role the Peach, but they are mistaken: for although it is good, that is, the well-chosen ones, but even better are several Pears, Apples and Plums, otherwise unworthy of being elevated to such a high degree. Returning to the Fig, I would say that the contemplation of such a fruit is admirable, for the abundant variety of its species and their various qualities, of which, as with Pears, the Ancients have spoken highly).[43]

In the tradition of poets who deal with things instead of abstractions, Serres would be an avatar of the author of *Comment une figue de paroles et pourquoi* ('How a fig of speech and why') or else, more famously, *Le Parti pris des choses* ('The Nature of Things', also known as 'Partisan of Things'), where for Francis Ponge mental consideration and physical tact are one and the same in the art of touching, naming, classifying, and cultivating an object.[44] The list (or *roole*) of figs has a faint military echo, named for a squadron of soldiers or a pair of combatants, defending and illustrating the cause of figs, who march in lockstep together. Variety—or biodiversity—endows the world with life. Serres notes:

[o]n en void des blanches, des noires, enfumées, grises, tannees, vertes. Des grandes, moiennes, petites. Des hastiues & tardiues. Des saueurs tres-diuerses & precieuses. Quand aux noms des Figues, il est à propos d'en représenter les plus vsités, & de mesme pour l'honneur de l'Antiquité, renouueller ceux qu'elle a voulu donner à ce bon fruit, que le temps n'a peu enterrer, non pour autre auantage, que pour contenter nostre curiosité. [...] . Es endroits de ce Roiaume où la Figue croist gaiement, on fait cas de celles qu'on nomme ainsi, Aubicons, Bourjaslotes, Blanquetes, Brunessenques, Quotidianes, Œil de perdix, Blauetes, Coucourolles, Bouueaux, Douces, Hospitalieres, Coquines, Roussaux, Pel-dure, Marseillettes, Angeliques, qui sont blanches, longues & grosses, Pourquines, noires & petites, pour brefueté obmettants les autres.

43 Ibid., pp. 697–698 and 1040. Emphasis mine.

44 Ponge, *Comment une figue de paroles et pourquoi* and *Le Parti pris des choses, precédé de douze petits écrits suivis des Proêmes*.

> (we see white ones, as well as black, smokey, gray, tan, and green ones. [There are] large ones, and medium and small ones. Fast growing, and slow. Flavours very diverse and precious. As to the name of Figs, it is right to represent the most common sorts, and also in honour of Antiquity, to bring back the names that [antiquity] wanted to give to this fine fruit, that time has been unable to bury, not for a specific advantage, but to please our curiosity [...]. In places in our Kingdom where the fig flourishes, we can mention those named thus: Aubicons, Bourjaslotes, Brunessenques, Quotidianes, Œil de perdrix, Blauettes, Coucourelles, Blanquettes, Boutueaux, Douces, Hospitalieres, Coquines, Rousseaux, Pel-dure, Marseillettes, Angeliques, which are white, long and fat, Pourquines, black and small, to be brief, in omitting the others).[45]

Curiosity drives the force of experience that received knowledge otherwise 'controls' [*contrerolle*]—Serres draws his knowledge here from Columella and other classical sources, and shows how artful arrangement (such as we find in the *Théâtre*) sets it before the reader's eyes to satisfy this curiosity.

Herein the geography of the nation intervenes. The author shows how much stronger *French* figs grow than those listed in classical sources. Serres first notes the places of this kingdom where the fig tree gaily grows. Attending to *terroirs* that figs enjoy, which the *manchette*, doubling the design of the book, calls 'Lieux de ce Roiaume esquels les bonnes Figures croissent' ('the Places in this Kingdom where good Figures grow'),[46] Serres includes his native space, Bourg Saint Andeol, 'ma patrie' (my homeland).[47] Can it be any coincidence that, as a result, the next species of tree—an *arbre* whose selection seems hardly *arbitraire*—inspires an association with the author's name? Like the fig that figures the warmth and temperance of the author's homeland, the olive tree—*olivier*—'tends more to the south than the north, liking better warmth than cold' and, as an emblem of productivity and worth, has no equal.[48] In fact, the author remarks, we *see* with the naked

45 Serres, *Théâtre*, pp. 698 and 1040–1041. Not all of the names are listed in Cotgrave. Among others, the Aubicon is considered one of the most delicious varieties, the Blanquete 'a delicate white summer peare', the Angelique, as Serres had noted, 'a kind of long, white, and great fig'. The variety attests to a commanding knowledge of both species and the local names that accompany them.

46 Ibid., pp. 698 and 1040.

47 Ibid., pp. 699 and 1042.

48 Ibid., pp. 701 and 1045. The text rehearses the contradiction of image and text in the inaugural chapter of Ferdinand de Saussure's *Cours de linguistique générale*, in which the image of an *arbre* is shown to have an 'arbritrary' relation with *arbre*, the corresponding word. Like Saussure, Serres 'motivates' the 'arbitrary' relation of words to things. The prose moves between the logics of analogy and representation that Michel Foucault describes at the outset of *Les Mots et les choses*,

eye (the *œil*) its fruit in its name: no tree 'ne le precede en valeur, pour la richesse qui prouient de son huile (*par excellence à ce seul mot, huile, estant recogneuë celle d'Olive*) & gentillesse de la confiture de ses Oliues' ('precedes it in value, because of the richness that comes from its oil (*for the excellence of this very word, oil, is recognized to be that of Olive*) and the kindness of the preserves of its Olives').[49] Furthermore, its green limbs not only survive the winter but also are sign of Peace: this coming two years after the signing of the Edict of Nantes, which the *Théâtre* promotes on the frontispiece.

The wealth, pleasure, and ecology of the compendium are found in the complex relation of description, agricultural history and figuration. The layout of the book implies that the reader should *see* the words as seeds bearing fruit when they are read and nurtured. The sixth *lieu,* devoted to the garden, becomes what might be the secret or even abyssal space in which the design of the work correlates its matter with a carefully designed display of map-like ground plans of gardens. The site, a sort of *hortus conclusus* close to the middle of the book, becomes one of measured and carefully managed growth. From the start, *mesnage*, a compass-word iterated everywhere in the 1,000 pages conveying the vision of the *Théâtre*, is associated with a prelapsarian garden in a postlapsarian time and space.

> Ce sont les Iardinages, qui fournissent à l'ornement utile de nostre Mesnage, innumerables especes de racines, d'herbes, de fleurs, de fruits, avec beaucoup de merueille. Aussi merueilleux en est le Createur, donnant à l'homme tant de sortes de viandes, differentes en matiere, figure, capacité, couleur, saveur, proprieté, qu'impossible est de le pouvoir toutes discerner ni comprendre.

> (These are the gardens that provide, for the useful ornament of our Household, innumerable sorts of roots, herbs, flowers, fruits, and with great wonder. No less wonderful is the Creator, giving to man all sorts of foods, so differing in material, figure, capacity, colour, taste, propriety, that it is impossible to be able to discern and understand them all).[50]

and it confirms Montaigne's remark at the beginning of 'De la gloire', where the word 'est une piece estrangere jointe à la chose' ('is a foreign morsel joined to the thing'). *Essais*, 2. 16, 697.

49 Serres, *Théâtre*, p. 701, emphasis mine.

50 Serres, *Théâtre*, pp. 499 and 781. The point is stressed over and again. Infinite variety comes from both without and within, especially in the medicinal garden, 'tant riche, qu'il est impossible de la representer entierement, à cause de l'infinie quantité de semences, racines, herbes, arbustes dont il est composé: n'estant mesmes entré en la cognoissance de l'homme, tout ce qui le peut rendre recommendable. Parce que de jour à autre, paroissent des nouuelles

Given the Creator's largesse and the ongoing discoveries of both local cultures and new plants (from foreign places *and* within the French kingdom), the project of a total representation, which would amount to containment, is impossible.

The beauty of variety depends on carefully designed gardens that enclose the fruits of nature in the confines of protected areas. There are sixteen woodcut images set within the text of the thirteenth and fourteenth chapters—'Emploi des herbes & fleurs, pour Bordures & Compartimens' ('The Use of Herbs and Flowers for Borders and Compartments') and 'Le Jardin Medicinal & son ordonnance' ('The Medicinal Garden and Its Organization') respectively.[51] The latter twelve, in the fourteenth chapter, are square compartments, five of which contain the insignia of Henry of Navarre. Some are virtual, others are taken from the legacy of recent Valois monarchs. Serres insists that those who walk along the pathways must discern the design from the ground—and also, thanks to the use of terraces, from above, from slightly elevated positions where the eye can see the space that extends while, almost paradoxically, remaining enclosed (since one compartment adjoins another).[52] Proportion and limit being difficult to calculate, only knowledgeable designers and inventive gardeners are up to the task of design and layout. Inasmuch as 'there are very few gardeners who have in their brain the inventions of exquisite compartments and other precious dispositions for the decoration of the pleasure',[53] the designer ought to follow the model of experts in portraiture: declaring 'ce sera de là d'où nos Jardiniers prendront leurs desseins' ('whence our gardeners will take their designs'), the author adds that such examples include the gardens the king had made at the Saint-Germain-en-Laye, the new Jardin des Tuileries, and Fontainebleau, all illustrated in the final pages of the chapter.[54] These twelve woodcuts make clear how the square and rectangle draw attention to the centres and edges of perceptibly contained areas. The compartments suggest that the spatial order can be used to celebrate the king where his device is set at the axis or corners, and that the extremity of one design

plantes estrangeres, voire des domestiques (incognues par nous le temps passé) paroissans pour y estre logees' ('so rich, because of the infinite quantities of sowings, roots, grasses, shrubs that comprise it: not even known to man, everything that recommends it. Because from one day to the next there appear new foreign plants, and even domestic species (unknown to us in times past) appearing so as to be included': *Théâtre*, pp. 598 and 906).

51 Ibid., pp. 598–605 and 905–912.

52 See also Le Dantec, *Poétique des jardins*, Chapter 1.

53 Serres, *Théâtre*, pp. 542 and 840.

54 Ibid., pp. 586–597 and 899.

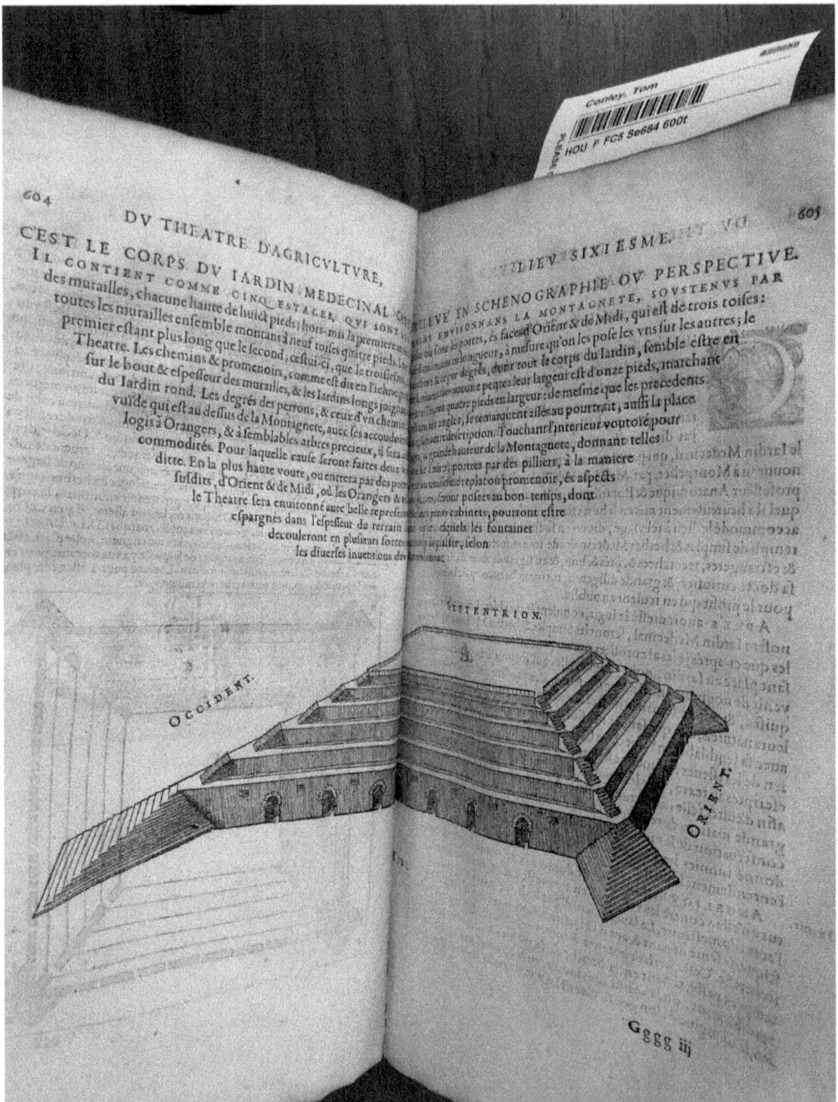

Figure 6: Hypothetical garden from the 'Sixième Lieu', from Olivier de Serres, *Le Théâtre d'agriculture et mesnage des champs*, pp. 604–605.

might serve as a centre for another (if indeed space allows for expansion).[55] The images celebrate the new design of the monarchic domains at the same time that, the text implies, they are both physical and mental templates for the real and mental gardener. They arrange space in ways analogous to the logic that informs the *Théâtre* as a whole.

In the sixteenth chapter, the medicinal garden (graced with four illustrations) seems to be more prominent than it is made in the rest of the text. Found in the great villas in Florence, Pisa, Padua, Genoa 'and other diverse places in Italy',[56] and associated with a few 'Princes and Republics' that have built them at great expense, the medicinal garden can nonetheless be designed within the *père de famille*'s more modest ambit. In reduced proportion its beauty and usefulness are no less commanding. Yet when many different species of plant are put together, Serres's experience with noblemen tells him that the practical garden requires a terraced, even pyramidal plan. A 'Montaignette' made from clay and sand, fertilized with manure, will offer different degrees of sun and exposure; its staggered levels or climates will allow different species to grow, and a pathway is indeed fashioned along its edge (hence the allusion to the Tower of Babel and the Tourré-magne of Nîmes).[57] It can be exposed and likewise surrounded with walls and protected from rain and cold. When set in a square, made with several paths and terraced alleys, it is 'like a Theatre', its steps giving way to each other, turning the garden into 'a very handsome representation' either with or without masonry, such that in all events a ground plan, diagram, or map remains the best way to imagine it.[58]

Four woodcuts follow. The first, of a circular design, appears to be an ichnographic view of an inverted theatre or an arena.[59] Its diameter measures forty-five *toises*, and it is built with a spiral pathway that can be followed 'tournoiant en volute' ('swirling in a curl') along the outside, leading to an open area (with a radius of ten *toises*) at the top. Four gateways, arranged according to the four cardinal directions, provide entry and egress. Along the pathway on the outside, garden-boxes are positioned so that the situation of each plant is determined by light and shadow or warmth and cold. The second

55 Edgerton sees in the quincunx plan the Roman heritage of conquest or strategic development that for Serres heraldic and self-contained. See his 'From Mental Matrix to *Mappamundi* to Christian Empire: The Heritage of Ptolemaic Cartography in the Renaissance'.

56 Serres, *Théâtre*, p. 598.

57 Ibid., pp. 600 and 910.

58 Ibid.

59 Ibid., pp. 601 and 911.

woodcut, a scenography of the 'Montaignette',[60] complements the ichnography. The author notes that the schematic design of the would-be tower does not permit detailed treatment of a balustrade or handrail along the spiral path. A fountain is placed at the centre of the circle at the top for ease when watering the boxes. The medicinal garden is strategically conceived. The pathways and four entrances that burrow into the edifice allow rapid deployment of workers; the top allows a commanding view of the entirety of the countryside. The conceptual architecture of the garden incorporates a further mode of defence where plants are grown to heal the sick, wounded, or frail.

The square design of a similar hypothetical garden resembling a ziggurat further emphasizes the military engineering. An ichnographic view displays the rows in the fashion of a terrace whose receding stages lead to a square promontory on the top, which also has a fountain at its centre. The bird's eye view of the fourth cut requires the printer to set the folio on a strip, an *onglet*, in the manner of a map bound in an Ortelian atlas (Fig. 4). Set in the virtual middle of the *Théâtre*, the image and its text become visual and textual mirrors of one another—and of the work itself, a building replete with gardens and a protective structure. A mental and physical fortress of the body of the garden 'appears to be a theatre' ('semble estre un Theatre'),[61] a specular image of what the title of the book had announced. The four *perrons* (sets of sixteen stairs) at each corner of the garden are suggestive of *courtines*, or angled battlements, which echo the evolving modes of construction to support new modes of warfare.[62] Yet although the garden resembles a *chateau fort*, it will also be sensuously arranged and decorated with 'Orangers & telles autres plantes' ('orange trees and other plants'). As if referring to the book itself, Serres adds, 'le Theatre sera enuironné avec une belle representation' ('the Theatre will be surrounded with a handsome representation').[63] The garden is designed to both guard and gratify. 'Plusieurs petits cabinets pourront estre espargnés dans l'espesseur du terrain sur chacun replat, esquels les fontaines decouleront en plusieurs sortes avec baucoup de plaisir, selon les diverses inventions des gens d'entendement' ('Several little compartments can be made from the thick ground on each of the [five] alley beds, in which fountains will flow in several fashions with a great deal of pleasure, according to the diverse inventions of informed and knowledgeable people').[64]

60 Ibid., pp. 602 and 912.

61 Ibid., pp. 604 and 912.

62 See David Buisseret, *Ingénieurs et fortifications avant Vauban: l'organisation d'un service royal aux XVI*ᵉ*–XVII*ᵉ *siècles*.

63 Serres, *Théâtre*, pp. 604–605 and 912.

64 Ibid.

The Closure

Last, but not least, *Le Théâtre* devotes its final 'place' to foodstuffs and to 'l'honneste Comportement en la Solitude de la Campagne' ('the honest way of living in the solitude of the country').[65] A virtual diagram (or familial flow chart) assigns to the father and mother the tasks of taking care of the household and maintaining the space and *habitus* for those who reside within, keeping medicine on hand to remedy humans and beasts, and finally enjoins them to 'Corriger la solitude de l'habitation Champestre' ('Correct the solitude of country living') through 'La Chasse, & autres honnestes exercices du Gentilhomme' ('Hunting and other honest exercises fitting for the gentleman').[66] Where much of the section (Chapters 1–6)[67] offers copious counsel and practical remedies for physical illness, the final pages address a commonplace in bucolic lyric: how to live in solitude. Hunting counts among the *provisions* that keep the mind at rest and the larder replete with freshly killed game. As seen elsewhere, the sport, exercise, and diversion that hunting provides carry echoes of war and—like the battery of medicines kept in the household—the need for self-protection. Venery is among 'utiles plaisirs' ('useful pleasures') that, like wine, is best practiced with moderation.[68] It rids the countryside of 'bestes ravissantes' ('ravishing beasts') who eat humans and livestock.[69] In times of peace it is, like Montaigne's *exercitation*, an art or practice rehearsing war. '[P]ar l'exercice de la Chasse, le Gentil-homme se façonne à la guerre, y apprenant les ruses de l'art, à s'endurcir au travauil, fuiant l'oisiveté, à se contenter de manger & boire peu, à s'accoustumer à toutes viandes & breuvages, à combatre à force & par surprinse, à piquer chevaux par bon & mauvais païs, dont il se rendra bon guerrier' ('By the practice of Hunting, the Gentle-man fashions himself for war, learns the stratagems of the art, to toughen up to labour, fleeing idleness, being content to eat and drink little, to be accustomed to all foods and beverages, to fight with force and by surprise, to ride horses in good and in bad country, which will turn him into a good soldier').[70]

Venery, Serres admits, cannot cure all ills. The prospect of living in solitude, which soon will haunt writers and philosophers of the century to follow, requires a philosophical programme that folds back upon what the

65 Ibid., pp. 815 and 1199.
66 Ibid., pp. 816 and 2000.
67 Ibid., pp. 817–991 and 2001–1442.
68 Ibid., pp. 992 and 1442.
69 Ibid.
70 Ibid., pp. 992–993 and 1443.

book has previously suggested in its account of its own genesis. In line with Montaigne's negotiation of his three 'commerces' (or kinds of association) in the third chapter of the third volume of the *Essais*, Serres asks his reader to consider the worth of reading. Unlike that of the essayist, however, the aim is not to get lost in the matter, nor even to change perspective or find utility in idle play. His ideal readers will use their time to hone their skills in logistics, engineering the management and defence of the land through consultation and cultivation of maps and the histories in their accompanying narratives:

> Scipion l'Africain […], disant à ses amis (qui s'esbahissoient de sa vie priuee & retirée) *n'estre jamais moins seul, que quand il estoit seul.* Si que le Gentil-homme aimant les livres, ne pourra estre que bien à son aise, auec vn livre au poingt se promenant par ses jardins, ses prairies, ses bois, tenant l'œil sur ses sens & affaires. En mauvais temps de froidures & de pluies, estant dans la maison, se promenera sous la guide de ses livres, par la terre, par la mer, par les Roiaumes & provinces plus loingtaines, *aiant les cartes deuant ses yeux, lui monstrant à l'œil leurs situations.* Dans l'histoire, contemplera les choses passées, les guerres, les batailles, la vie & les mœurs des Rois & Princes, pour imiter les bons, & fuir les mauvais. Remarquera les gouuernemens des peuples, leurs loix, leurs polices, leurs coustumes, tant pour entendre comme le monde se gouverne, que pour faire profit des salutaires auis qu'il en pourra tirer, les appropriant à ses usages.
>
> (Scipio Africanus […] , telling his friends (who would be astounded by his private and retired life) *that he was never less alone than when he was alone.* So much so that the Gentle-man who loves his books, will only be at ease with a book in hand, meandering through his gardens, his fields, his woods, keeping an eye over his senses and affairs. In bad weather, of cold and rain, in the confines of home, will meander guided by his books, through lands and seas, through Kingdoms and provinces more distant, *by having the maps under his eyes, showing him their situations.* In history, he will contemplate things past, wars, battles, the lives of Kings and Princes, in order to imitate the good and shun the bad. He will notice the governments of the peoples, their laws, their politics and their customs, not only to understand how the world is governed, but to profit from the salutary opinions he could gain by appropriating them for his own uses).[71]

71 Ibid., pp. 996 and 1448, emphasis mine.

A familial counterpart to Machiavelli's prince, who is told to get the lay of his land, Serres's solitary gentleman will use books to situate himself and to turn his country estate into a well defended and exquisitely managed territory. To be sure, he anticipates Rousseau's *promeneur solitaire*, who identifies the herbs and medicinal plants growing about him, but to a strong degree he becomes an analogue of the king's engineer, the surveyor and cartographer assigned to redesign and to reshape the borders of the nation. As if having read his Peter Apian or Antoine du Pinet, he will make strategic use of 'l'Arithmetique, la Geometrie, l'Architecture, la Perspectiue, mesme la Pourtraiture, pour representer forteresses, villes, chasteaux, païsages, dignes parties du Gentil-homme, moiennent lesquelles, il desseignera plans de forteresses, & de maisons privees, voire par tels moiens, ordonnera de ses bastimens, de ses jardins, de la disposition de ses arbres, & fera autres choses de son mesnage par art, avec heureuse issue' ('Arithmetic, Geometry, Architecture, Perspective, even Portraiture to represent fortresses, cities, chateaus, landscapes, all worthy of the Gentleman, by means of which he can design the plans for both fortresses and private houses, indeed by whose means he can bring order to his buildings, to his gardens, to the arrangement of his trees, and do other things of consequence for his dwelling through his own craft'). In his solitude, far from falling into melancholy or sloth, the gentleman and ideal reader of the *Théâtre* becomes both a strategist and an artist.

Along these lines of solitude and self-management, the conclusion of the *Théâtre* combines the topics it explores with its own mode of production and its own particular form. The pastoral patriarch's governance of his farm and fields has as proof its allusion to the book-as-property or the well-kept estate in paper and ink. In the instances where Serres tenders the analogy between the order of the book and how it 'represents' the components of his theatre, the analogy prevails. The book he writes becomes his 'field' of reference. The incipit to the conclusion could not be clearer:

> Des paroles il faut venir aux Effets, pour auoir contentement de nostre Agriculture. Et comme ce n'est que du papier peint, que le dessein du bastiment, sans pierre, chaux, sablon, bois, & autres materiaux, pour esleuer l'edifice, aussi vainement aurions-nous representé le Mesnage des champs, sans mettre la main à l'oeuvre. L'on accoutumé de se moquer de ceux qui disent vouloir bastir, planter, reparer, sans en voir l'auancement. Voire les terres mesmes, semblent accuser de negligence leur posesseurs, qui ne les mettent en poinct d'enfanter les biens qu'elles ont conceu dans les entrailles.

(We need to turn words into Effects, in order to be satisified with our Agriculture. And since it is nothing but painted paper, nothing but the drawing of the building, without stone, lime, sand, wood or other materials to elevate the edifice, we would have represented the management of fields in vain, without putting a hand to the labour. We have the custom of mocking those who say they want to build, plant, repair, without takin note of progress. The lands themselves seem to accuse their owners of negligence, who do not allow them to grow from the seeds that they have produced in their entrails).[72]

The 'work' to which the author has put his callused hands is *Le Théâtre*, and as such its own form—its disposition, its order, arrangement, balance of classical wisdom and lessons drawn from experience—carries much of its mission to 'represent' the matter of the landowner's labours.

The history or vision that planted the seeds of the book is found in reference to Xenophon's Cyrus, who 'estimoit les plus belles occupations du Gentil homme estre de l'agriculture et la guerre, luy mesme s'emploiant & en l'un & en l'autre exercice' ('considered the most beautiful occupations of the Gentle-man to be agriculture and war, himself being useful in both exercises').[73] To be sure, in their passage from a state of war to peace, swords are turned to pruning hooks, but only such that the virtues of the one are rooted in the other. War, which recedes from the horizon in 1600, is not something the sight of well-managed fields and farms causes populations to forget. The formulation showing that the one is *in* the other anticipates Cartesian effects where peace and solitude are found not in retreat from war but, less obviously, in the knowledge that the memory of strife is embedded in the soil. Among other ancients, Cyrus took pleasure in rewarding his provincial deputies with land in return for their good upkeep of the earth, proper distribution of local population, and wise planting of trees bearing fruit. By contrast, he redistributed the lands of those who had let them go to seed or fail to populate them as they ought. The best subjects are those managing small plots with *good reason* and without the ambition of gaining public office. Furthermore, the rustic and *solitary* life led with modest means, as Serres notes with regard to Plato in the *Republic*, were 'the patron and example [*maistresse & exemple*]' of all sobriety, continence, parsimony, and diligence. In praise of the solitary place, the descriptive topography depicts an assessor looking at a landscape he seeks to protect,

72 Ibid., pp. 998 and 1449.

73 Ibid., pp. 999 and 1450.

wishing 'de ne changer ou arracher les bornes des champs, de ne romper les canaux, ou destourner le cours des eaux qui abruuent les terres, de ne faires degasts ax heritages, ni aux fruits de la terre, sous grandes peines' ('not to change or tear apart the borders of fields, break the canals, or deviate the courses of rivers that water the lands, let heritage or the fruits of the earth be devastated without heavy fines').[74]

It is here where the concept of *Théâtre*, the key spatial component of the project, returns as an element of a greater landscape. An analogy carries the figure into the agrarian world: '[T]out ainsi, que les grandes & superbes villes & cites, servent de Theatre & de spectacle à nos miseres & calamités, ainsi les champs solitaires, couvrent nos imperfections & infirmités, toutes choses honnestes y estans receuës, quoi-que de peu de lustre' ('Just as the great and superb towns and cities act as Theatre and spectacle of/for our miseries and calamities, so the lonely fields cover our imperfections and infirmities, all honest things being received there, though of little brilliance').[75] When he goes to the city, the good *mesnager* is obliged to turn his 'free' fashion of living into its 'servile' antithesis and convert 'his peace into labour [*son repos, en trauail*]'. In the virtual architecture of the book, the closing reflections on productive solitude are an echo of the seemingly autobiographical reflection that inaugurated the project. War had forced the author to take refuge in both his country home and in himself, and to consider—by way of analogy—that, in the way it was managed, his home and family (*maison*) had the good fortune of being more a site of peace than of war. Solitude, implied by the admission that the author was *chez soy*, at home with himself while collaborating with hired hands, gave rise to the project:

> Mon inclination, & l'estat de mes affaires, m'ont retenu aux champs en ma maison, & fait passer une bonne partie de mes meilleurs ans, Durant les guerres civiles de ce Roiaume, cultivant la terre par mes serviteurs, comme le temps l'a peu porter. En quoi Dieu m'a tellement comporté parmi les diverses humeurs de ma Patrie, que ma maison aiant esté plus logis de paix que de guerre, quand les occasions s'en sont presentees, j'ai rapporté ce tesmoignage de mes voisins, qu'en me conservant auec eux, je me suis princiaplement addonné chés moi, à faire mon mesnage.
>
> (My own inclination and the state of my affairs have retained me in the fields in my home, and have made me spend a good part of my best years,

74 Ibid., pp. 999–1000 and 1452.

75 Ibid., p. 1002.

> during the civil wars of this Kingdom, cultivating the earth through those who serve me, as time has shown. In which God has included me in the various complexions of my Homeland, such that my home has been more a house of peace than of war, [and] whenever the occasion was offered, I brought this testimony of my neighbours, that by conserving myself with them, I principally devoted myself, in my home, to holding to my business).[76]

In times of calamity and in the give-and-take of war and peace, in the midst of uncertainty and occasional calamity, care of an estate, which in this essay is taken to be a form of ecology, becomes a meditation on an inner space that the author 'manages' through reading and writing. The solitude praised at the end of the volume underscores how the science of the book pressed upon the imperiously personal need to write:

> Durant ce misérable temps-là, à quoi eussé-je peu mieux emploier mon esprit, qu'à rechercher ce qui est de mon humeur? Soit donc que la paix nous donnast quelque relasche, soit que la guerre par diverses recheutes, m'imposast la necessité de garder ma maison, & les calamités publiques, me fissent cercher quelque remède contre l'ennui, trompant le temps, j'ai treuué un singulier contentement, âpres la doctrine salutaire de mon ame, en les lectures des liures d'Agriculture, à laquelle j'ai de surcroist ajousté le jugement de ma propre experience. Je dirai donc librement, qu'aiant souvent & soigneusement leu les livres d'Agriculture, tant anciens que modernes, & par expérience observé quelques choses qui ne l'ont encores esté, que je sache, il m'a semblé estre de mon deuoir, de les communiquer au public, pour contribuer selon moi, au vivre des hommes. C'est ce qui m'a fait escrire.

> (During this miserable time, how could I have better employed my mind than searching what was of my own disposition? Whether the peace gave us some rest, or whether the war by various relapses forced me to remain at home, and the public calamities made me look for a remedy against boredom, to pass the time I found a singular contentment, after the salutary doctrine of my soul, in reading books of Agriculture, to which I have added the judgment of my own experience. I will therefore say freely, that having often and carefully read books of Agriculture, both ancient and modern, and by experience having observed some things that have

76 Ibid., fol. ãi v and p. 60.

not yet been accounted for, to my knowledge, I found it to be my duty to communicate them to the public, to contribute, in my opinion, to the lives of men. That is what made me write).[77]

Getting lost and finding oneself is a matter that the book takes up frequently, both in its form and its treatment of content. We are led into the past, where the author deals with the knowledge drawn from classical sources, and carried into the present with reports of what is observed about practices on French soil. The text becomes the forest whose compass points that locate the reader are the *manchettes*, images, diagrams, intercalated proverbs, and a copious index. In short, and in abrupt conclusion, when displaced into our moment, the economy that goes with the concept and practice of *mesnage* has the tenor of a practical ecology.

Works Cited

Balzac, Honoré de, *La Peau de chagrin*, ed. by Nadine Satiat (Paris: Garnier/Flammarion, 1996).

Buisseret, David, *Ingénieurs et fortifications avant Vauban: l'organisation d'un service royal aux XVIe-XVIIe siècles* (Paris: CHTS, 2002).

Conley, Tom, *An Errant Eye: Poetry and Topography in Early Modern France* (Minneapolis: University of Minnesota Press, 2011).

Cotgrave, Randle, *A Dictionarie of the French and English Tongues* (London: Adam Islip, 1611).

Delaune, Étienne, *Douze Mois* (Paris: s.n., 1560). [Houghton Library, Harvard University call number: *FC5.D3755.560l].

Edgerton Jr., Samuel Y., 'From Mental Matrix to *Mappamundi* to Christian Empire: The Heritage of Ptolemaic Cartography in the Renaissance', in *Art and Cartography: Six Historical Essays*, ed. David Woodward (Chicago: University of Chicago Press, 1987), pp. 10–50.

Foucault, Michel, *Les Mots et les choses* (Paris: Gallimard, 1966).

Goul, Pauline, 'The Vanity of Ecology: Expenditure in Montaigne's Vision of the World World', in *French Ecocriticism. From the Early Modern Period to the Twenty-First Century*, ed. by Daniel A.Finch-Race and Stephanie Posthumus (Berlin: Peter Lang, 2017), pp. 43–57.

Gourdin, Henri, *Olivier de Serres: 'Science, experience, diligence' en agriculture au temps d'Henri IV* (Arles: Actes Sud, 2001).

77 Ibid.

Holbein, Hans, *Les simulachres & historiees faces de la mort, avtant elegammẽt pourtraictes, que artificiellement imagines* (Lyon: Soubz l'escu de Cologne, 1538). [Houghton Library, Harvard University call number: Typ 515.38.456].
Lavondès, Antoinette, 'Les éditions du "Théâtre d'agriculture"', *Bulletin de la société de l'histoire du protestantisme français* 88 (1939), 160–162.
Le Dantec, Pierre-Georges, *Poétique des jardins* (Paris: Actes Sud, 2011).
Machiavelli, *Le Prince de Nicolas Machiavelle secretaire et citoien de Florence*, trans. by Guillaume Cappel (Paris: Chez Charles Estienne Imprimeur du Roy, 1553).
Montaigne, Michel de, *Essais*, ed. by Pierre Villey and V.-L. Saulnier (Paris: PUF, 1978).
———, *The Complete Essays*, trans. by Donald Frame (Stanford: Stanford University Press, 1958).
Mortimer, Ruth, *Catalogue of Books and Manuscripts* (Cambridge: Belknap Press of Harvard University, 1964).
Mukerji, Chandra, *Territorial Ambitions and the Gardens of Versailles* (Cambridge: Cambridge University Press, 1997).
Ponge, Francis, *Comment une figue de paroles et pourquoi* (Paris: Flammarion, 1977).
———, *Le Parti pris des choses, precédé de douze petits écrits suivis des Proêmes* (Paris: Gallimard, 1967).
Serres, Olivier de, *La cueillete de la soye, par la nourriture des vers qui la font: Echantillon du Theatre d'agriculture d'Olivier de Serres seigneur du Pradel* (Paris: Chez Jamet Mettayer, Imprimeur ordinaire du roy, 1599). [Houghton Libraray, Harvard University call number: *FC Se684 599c].
———, *Le Théâtre d'agriculture et mesnage des champs* (Paris: Jamet Mettayer, 1600).
———, *Olivier de Serres, père de l'Agriculture française: Le Théâtre d'Agriculture et mesnage des Champs*. Pages choisies, précédées d'une lettre de M. le Maréchal Pétain (Paris: Firmin-Didot et Cie., 1941).
———, *Le Théâtre d'agriculture et mesnage des champs*, ed. by Pierre Lieutaghi (Arles: Actes Sud, 2001).
Usher, Phillip John, *Exterranean: Extraction in the Humanist Anthropocene* (New York: Fordham University Press, 2019).
Zerner, Henri, *L'Art de la Renaissance en France: L'Invention du classicism* (Paris: Flammarion, 2002).

About the Author

Tom Conley (Abbot Lawrence Lowell Professor of Visual and Environmental Studies and of Romance Languages and Literatures, Harvard University) studies relations of space and writing in literature, cartography, and cinema. His work moves to and from early modern France and issues in theory and interpretation in visual media. His books include *Film Hieroglyphs* (1991, new edition 2006), *The Graphic Unconscious in Early Modern Writing* (1992), *The Self-Made Map: Cartographic Writing in Early Modern France* (1996, new edition 2010), *L'Inconscient graphique: Essai sur la lettre à la Renaissance* (2000), *Cartographic Cinema* (2007), *An Errant Eye: Topography and Poetry in Early Modern France* (2011), and *À fleur de page: Voir et lire le texte de la Renaissance* (2014). His translations include Michel de Certeau's *The Writing of History* (1988 and 1992) and the same author's *Capture of Speech* (1997) and *Culture in the Plural* (1997), Marc Augé's *In the Metro* (2003) and *Casablanca: Movies and Memory* (2009), Gilles Deleuze's *The Fold: Leibniz and the Baroque* (1993), and Christian Jacob's *The Sovereign Map* (2006), among others. He is also the author of over 250 articles and book chapters. He has held visiting appointments at the University of California, Berkeley, UCLA, the Graduate Center of the City University of New York, L'École de Chartes, L'École des Hautes Etudes en Sciences Sociales, and other institutions. In 2003 he was a seminar leader at the School for Critical Theory (Cornell). He has been awarded fellowships from the American Council for Learned Societies, the National Endowment for the Humanities, and the Guggenheim Foundation.

11. Montaigne's Plants in Movement

Antónia Szabari

Abstract

By zooming in on the diverse sources of Montaigne's naturalism, from Aristotelian notions of vegetal *psyche* to Epicurean atomism and everyday observation, the essay examines the figures of plants in Michel de Montaigne's *Essays*. It reveals the remarkable animation that characterizes Montaigne's plants and argues that the essayist viewed not only animals (as has already been argued) but also plants are analogous to human beings and as forming the basis for moral judgments.

Keywords: vegetal ontology, naturalism, materialism, early modern ethics, Epicureanism, Montaigne's *Essays*

Introduction[1]

Animals tend to eclipse plants in our sightings of nonhuman others in Michel de Montaigne's *Essays* (*Les Essais*, 1580–1593).[2] In her magisterial book on libertine botany, *Quand l'esprit vient aux plantes*, Dominique Brancher explains the centrality of the animal in Montaigne by showing that Montaigne's sceptical turn granted animals different but equally valid perceptions, putting into question the conventional ontological and religious hierarchy

1 It was in the academic year 2000/1, in the context of a graduate seminar held by the late Gérard Defaux at Johns Hopkins University, that I first became aware of Montaigne's ethical regard for plants. My current return to these vegetal figures of thought in Montaigne, however, owes much, indeed more than it is possible to acknowledge here, to a book-length project, co-written with Natania Meeker and entitled *Radical Botany: Plants and Speculative Fiction*, forthcoming from Fordham University Press in 2019.

2 For example, there is no mention of plants in the *Dictionnaire de Michel de Montaigne*, ed. Philippe Desan. This article makes the case for including plants as important figures of thought in Montaigne's oeuvre and asks how doing so would change our interpretation of the book.

Goul, P. and P.J. Usher (eds.), *Early Modern Écologies. Beyond English Ecocriticism.*
Taylor & Francis Group, 2020
DOI 10.5117/9789462985971_CH11

that privileges the human.[3] Libertine botany, the subject of Brancher's book, subsequently takes this enquiry into the garden, where we find this destabilizing and multiplying of perspective in Cyrano de Bergerac's rational cabbage or Guy de la Brosse's affectionate herbs.[4] But what about the plants that crop up in *The Essays*? Did Montaigne's thought remain zoocentric, as Brancher claims, or does an attentive reading reveal that plants teach us something vital about Montaigne's relation to the natural world, to the self, or to the practice of writing? Our first clue is both resounding and enigmatic: in the essay 'De la cruauté' (II. 11), Montaigne asserts 'Quand tout cela en seroit à dire, si y a-il un certain respect qui nous attache, et un general devoir d'humanité, non aux bestes seulement qui ont vie et sentiment, mais aux arbres mesmes et aux plantes' ('Even if all of that remained unsaid, there is a kind of respect and a duty in a man as a genius which link us not merely to the beasts, which have life and feelings, but even to trees and plants').[5] What concerns me in this article is the possible significance of such a claim about our ethical obligations to plants, given Montaigne's unwillingness to return to it in any clear or explicit way elsewhere in his work. The reader would search in vain through the nonhuman examplars in the 'Apologie de Raimond Sebond' (II.12) that serve to show the similarity of so-called 'lower' creatures to the human being: not a single one is devoted to plants. While the author famously wondered about the intentions of his cat ('ma chatte'), the author claims to have very little knowledge of physical plants, even those that grow on his estate, let alone (as Brancher points out) the notion of attributing feelings or thoughts to them. Yet this unique mention of a human 'duty' toward plants is echoed by countless passages that present plant life in terms that allow us to outline the reasons why Montaigne may want us to turn to it with respect.

Even Montaigne seems to restrict his ethical claim to the following sentence: 'Nous devons la justice aux hommes, et la grace et la benignité aux autres creatures qui en peuvent estre capables' ('We owe justice to men:

3 Dominique Brancher, *Quand l'esprit vient aux plantes: botanique sensible et subversion libertine* (*XVII–XVIIIe siècles*).

4 Even when it accomplishes this decentralization of the human perspective, the project of libertine botany, Brancher argues, remains anthropological. It is worth noting that today anthropology is turning toward nonhuman others (including animals and plants) for an even greater decentering of the human perspective. On plants as anthropological subjects, see the works of Natasha Myers, for example, her essay 'Photosynthetic Mattering: Rooting into the Planthroposcene' in *Moving Plants*, ed. Thorsten, pp. 123–127.

5 Montaigne, *Les Essais*, ed. Villey, 435a; Montaigne, *The Complete Essays*, trans. Screech, p. 488. Hereafter, all parenthetical references in the body of the article will refer to there two editions, respectively.

and to the other creatures who are able to receive them we owe gentleness and kindness').[6] He does not include plants in the category of beings with sentiment; therefore, he does not ask us to treat them with acts of kindness.[7] However, I argue that, while Montaigne does not apply the Pyrrhonic turn (so aptly analysed by Brancher) to plants, who thus do not gain their own perspective, rational thoughts, or emotions, nonetheless a sense of familiary mixed with wonder does apply to plants in the *Essays*. If animals have dominated our study of nonhuman others, and the body has been seen as the primary sight of Montaigne's Epicureanism, then his turn to plants (to riff on the 'plant turn' that we have seen in contemporary theory, anthropology, philosophy, and the arts)[8] reveals the much greater degree to which Montaigne was willing to re-evaluate the human and cultural world by placing them in a natural and material context.

I begin by observing the strikingly animated character of plants, who shed their Aristotelian–Scholastic sedentariness. Here, Montaigne takes his cues from a work of medieval natural theology that he translated to allow him to materialize Aristotelian and theological conceptions of plants. Another important source for the sheer abundance of plants that appear in many images in the *Essays* is the Epicurean movement of matter. Does Montaigne talk about actual plants or are plants relegated to the figurative layer of the text, as ornaments or grotesques? Do his plant images merely form a figural layer contained in and by the project of self-portraiture, or does his fecund thought about plants influence and shape his ethical thinking and push it beyond zoo- and anthropocentrism? My title serves as a playful reference to and a modification of the title of one of the most notable classical analyses of Montaigne's powers of self-representation.[9] In this chapter I rely on the work of scholars who have studied Montaigne's unique breed of naturalism and his non-dogmatic Epicureanism over the past decade, while evoking contemporary new materialist critic Jane Bennett's notion of 'vibrant matter'. I argue that Montaigne's ethical injunction

6 Montaigne, *Les Essais*, 435a / *The Complete Essays*, p. 488.

7 Thus, facing plants, we do not find ourselves in conundrums of appetite and love, as does the lady who is told that there was a cat in the pie she had consumed in 'De la force de l'imagination' (I. 21). See Shannon, 'La chatte de Montaigne'.

8 On this notion of 'plant turn', I will limit myself to mentioning Jeffrey T. Nealon's *Plant Theory*, in which he zooms in on the incompleteness of plants already present in the Aristotelian notion of vegetal *psyche* and makes it into a positive concept as atelic life, marginally present in the works of Michel Foucault and Jacques Derrida and more central in those of Gilles Deleuze and Félix Guattari.

9 Starobinski, *Montaigne en mouvement*.

that we should respect plants is to be taken seriously, and analyse some of the ways in which plants matter in the *Essays*. By uncovering botanical thought and writing in the *Essays*, we discover Montaigne's deep commitment to the material world as well as to thinking and writing as immanent practices therein, which together reveal a materialism in which matter does not preceed thinking or representation as an inert object of human manipulation; nor does its movement follow a causal logic, as in the emerging forms of proto-scientific thought.

Montaigne's Overgrown Epicurean Garden

Montaigne was a rural gentleman who managed his estate alongside his wife. As Ullrich Langer notes, 'after 1570 the essayist was in touch on a daily basis with the peasants working his lands and the people employed on his estate'; on the other hand, there is not much direct reflection on this activity, on agriculture, or on the production of crops or wine in the text of the *Essays*.[10] With the nonchalance one might expect from the scion of an upwardly mobile, socially ambitious noble family of the middling sort, who wants to distance himself from his lands—his primary source of revenues—he states: 'Je suis né et nourry aux champs et parmy le labourage; j'ay des affaires et du mesnage en main, depuis que ceux qui me devançoient en la possession des biens que je jouys, m'ont quitté leur place. Or je [...] ne sçay la difference de l'un grain à l'autre, ny en la terre, ny au grenier, si elle n'est par trop apparente, ny à peine celle d'entre les choux et les laictues de mon jardin' ('I was born and brought up in the country, surrounded by agriculture; farming and its concerns have been in my hands ever since those who previously owned the lands which I enjoy moved over for me; yet [...] unless it is all too obvious I do not know the difference between one grain and another, neither in the ground nor in the barn; and in my vegetable garden I can scarcely tell cabbages from my lettuces').[11] Following his father's model (who did not cultivate the lands himself but leased them), Montaigne fashioned himself as an educated nobleman, but he did return to the management of his estate.

The second half of the sixteenth century saw political unrest, economic pressures due to demographic growth, crop failures and epidemics, and food scarcity. Inflation caused the price of grain to go up, and wars made

10 Langer, 'Montaigne's Political and Religious Context', p. 20.

11 Montaigne, *Les Essais*, 652 / *The Complete Essays*, p. 741.

estate owners—the 'colonists' of the countryside,[12] as Tom Conley puts it—vital. Judging from the brief account he gives of his finances, Montaigne did fairly well in this economy, which presented opportunity for some and impoverishment for many. But from this social historical context, I would like to move now to the passages that acknowledge his ties to his lands, and especially to plants, which figure often in the *Essays*. Even his disclaimer of agronomic expertise should evoke for us the world of agriculture, fields and landscapes, a site of ignorance which is nevertheless not a sign of failure but, rather, a starting point. The image Montaigne presents of himself, his failure at good management, cannot be fully accounted for (as, for example, Philippe Desan has argued) by his superficial or temporary appreciation of the countryside and the estate from which he hurries back to books, the company of men (and women), and the political world of his ambitions; however, it does offer the possibility for the transformation of the familiar agricultural world into a less familiar one.[13]

Montaigne's botanical images appear to be informed not only by his readings of philosophical and literary texts (Aristotle, Scholastic philosophy, Lucretius, Plutarch, and Ovid) but also by an interest in botany, the medicinal uses and proto-scientific observation of plants, and—last but not least—by the ordinary availability of plants for observation in the environment. There are brief but unmistakable hints to support this, including Montaigne's account in his journal of learning from a woodworker and maker of mathematical instruments in Pisa that tree rings denote the age and position of the tree[14]—an insight into plant physiology that may have originated in Leonardo da Vinci's tree studies. Similarly, his suggestion that young women should not be barred from studying botany, even when it teaches them about sexuality, is an argument that would be played out frequently, more often than not in Montaigne's favour, in the ensuing centuries.[15] Montaigne recognized and was interested in (and to some

12 Conley, 'Civil War and French Better Homes and Gardens', pp. 725–759.

13 Desan, *Montaigne: A Life*.

14 See the entry for Saturday, 8 July 1580. Michel de Montaigne, *Journal de Voyage*, pp. 191–192. See also Leonardo da Vinci, *Trattato della Pittura*, p. 396.

This insight by the Italian artist and polymath was largely overlooked by Montaigne's contemporaries. The two seventeenth-century natural historians Marcello Malpighi and Nehemiah Grew (the latter in his *Anatomy of plants*) are the first to mention it in print.

15 The anecdote relates Montaigne's daughter reading a French book during her botany lesson and being abruptly stopped by her governess when coming across the word '*fouteau*', the name of a tree ('beech') that by sound evokes a vulgar expression in French, 856b / 966. See Tom Conley's 'Montaigne moqueur: "Virgile" and Its Geographies of Gender' in *High Anxiety: Masculinity in*

degree admitted to experimenting with) the medicinal powers of plants.[16] These instances are in step with the nascent naturalism that characterizes the medical professions at the time, and involve a naturalization of plants (along with the larger natural world) which show Montaigne was thinking in line with other naturalists such as Amboise Paré. As Brancher shows, sixteenth-century botany was not particularly invested in questioning the place accorded to plants in an ontological hierarchy, and the knowledge sought about plants largely concerned their medicinal use. Studies in plant physiology did not take hold until the seventeenth century, with Guy de la Brosse's pioneering observations of plant life and the rise of zoophytes, which later in the eighteenth century led to the wholesale questioning of the status of plants vis-à-vis animals. There is more to this naturalization of plants, however, than simply gearing the knowledge about them towards a scientific view of the world, which makes the intellectual treatment of plants in the sixteenth century in general, and in Montaigne's thought in particular, especially noteworthy.

Although Montaigne lacked insights into plant physiology (apart from the exceptional case of the tree's rings), in his writings he shows an interest in the 'aliveness' of plants on several occasions. In 'De la cruauté' (II. 11), Montaigne suggests that punishment be aimed 'contre l'escorce, non contre le vif' ('at the dead bark not the living tree').[17] This is the essay in which the author makes the strongest claim about our ethical obligations toward plants, and here the 'living' part of the tree is figured as the plant pulp, as opposed to the dead bark. This statement precedes Montaigne's final conclusion about the ethical obligations that we owe to plants (as well as animals); while he does not attribute sensation to the tree, he does attribute life, which allows him to highlight the similarities between trees and human beings. Were the trees here simply an allegory for the more complex sufferings of human beings subjected to torture, this would be an anthropocentric argument; however, something else motivates Montaigne to evoke the living tree, including perhaps the memory of the Pisan woodworker's explanation of how the tree's rings indicate *le vif.* This image is one of the many instances in the book in which plants and human beings are viewed as similar, the similarity quite often functioning as a recognition of a palpable relation,

Crisis in Early Modern France, ed. Long, pp. 93–106. Conley's article also sheds light on some of Montaigne's most powerful vegetal images.

16 Montaigne seems to believe in those 'homely simples' (767a / p. 867, 772a / p. 872,) and he does not doubt the effect of plants such as horse-radish, senna pods (772a / p. 872).

17 Montaigne, *Les Essais*, 432b / *The Complete* Essays, p. 484.

like the sharing of a 'flesh' and a 'shell' or 'skin', rather than as a figure in which the nonhuman simply stands for the human.[18]

A possible philosophical source for Montaigne's botanical thought and imagery is the *Natural Theology* written by the Medieval Catalan scholar Raymond of Sabunde (d. 1436), which Montaigne had translated (as he says) to please his dying father. Sabunde conventionally places plants on the 'second degree' of the great chain of being (second from the bottom, after rocks) and describes them as having 'being and life only'; less conventionally, he also notes (in Montaigne's translation, which is telling) that they are defined by a capacity for *autonomous* movement into all directions of space:

> La seconde marche de nostre echelle comprend toutes choses qui ont l'estre et le vivre seulement: et dit-on qu'elles ont vie, d'autant que de soy elles se mouvent contrement, contrebas, devant, derriere, à dextre et à senestre: là sont toutes les plantes, les arbres et les herbes, qui vivent, d'autant qu'elles ont ce movement par elles mesmes. Nous les voyons croistre en hauteur et en grosseur, et tirer de la terre leur nourriture, par laquelle continuellement elles s'augmentent, s'entretiennent, engendrent de la semence et du fruict.
>
> (The second degree of our ladder comprehends all things that possess only being and life. It is said that they have life insofar as they move by themselves upward and downward, forward and backward, to the right and to the left. Here are all the plants, trees, and grasses that are alive insofar as they move by themselves. We can see them grow in length and thickness and draw their nutrition from the soil, with the help of which they constantly augment their size, stay alive, and engender fruit and seeds).[19]

For Aristotle—whose concept of plant 'soul' (*psūchē*) Sabunde interprets here, in Montaigne's translation—plants had limited capacity for growth, decay, and self-nurture.[20] Their 'soul' was in fact a rational form, impressed

18 Nor is there an analogy between plants and human beings, which will be the tool of eighteenth-century scientists studying plants: most notably Julien Offray de La Mettrie, possible author of *L'homme plante*, 1748, originally published anonymously in Germany, with a citation from Ovid's metamorphosis describing Daphne's transformation into a Laurel tree.

19 *Oeuvres complètes de Michel de Montaigne*, ed. Armaingraud, vol. 9: 'La Théologie naturelle de Raymond Sebon, pp. 7–8. The translation is mine.

20 For a critique of the concept of plants as privative in Aristotle in particular and metaphysics in general, see Michael Marder's *Plant-Thinking: A Philosophy of Vegetal Life.*

upon matter, and according to the model of nested souls this was equally present inside animals and human beings. Scholastic theology, by and large, found the nested model unacceptable and transformed it into a scale, where the sole rational (and thus immortal) soul belonged to human beings; consideration of all other 'souls', including much of human psychology, tended to be relegated to physics.

However, if, as Brancher argues, the theological appropriation of the vegetal soul and its placement at the bottom of a hierarchy are moves that tend to denaturalize plants, Montaigne follows in Sabunde's footsteps and naturalizes plant life. Although plants were assigned a limited capacity for growth and reproduction by Aristotle (and in Scholastic theology), they gain an excessive autonomy in Sabunde's description. This emphasis the Catalan theologian places on the movement of plants may have appealed to Montaigne because, if taken outside its context and teleological frame (i.e. the great chain of being), this movement of plants *par elles memes* (by themselves) seems to embody the autonomous movement of matter in general, an idea that Montiagne found in much more vivid images in Lucretius. This passage from Montaigne's translation of the medieval theologian's work (anathema to Roman authorities because it presented nature as a site of revelation) anticipates the way in which the essayist treats plants. Montaigne's plants are all movement—far from mostly lacking qualities, they offer a potential revelation about physics, nature, and the human psyche and body, too, insofar as the latter two exceed culture and require taming.

Michel Jeanneret has shown that Montaigne's fascination with the philosophy of becoming *naître* (being born), rather than *être* (being) draws on antique sources such as Plutarch (in Amyot's translation) and Epicurean doctrines, as described in Lucretius's *De rerum natura.*[21] Plants do indeed embody this perpetual motion in Montaigne's writing. Thus, while animals lead Montaigne into an anthropological investigation, plants are channels into matter, physics, the observable, the intimate, and the cosmic material world. Montaigne's references to plants seem to carry a conviction of their aliveness insofar as they become mediums or mirrors in which the cosmos in flux is revealed to the observer. Jeanneret asks whether the fluctuation of matter with which Montaigne was fascinated might ultimately, as in

21 Michel Jeanneret, *Perpetual Motion: Transforming Shapes in the Renaissance from da Vinci to Montaigne*. Originally published in French under the title *Perpetuum mobile: Métamorphoses des corps et des œuvres de Vinci à Montaigne*.

an Aristotelian ontology, aspire to form.[22] He responds by signalling the inadequacy of the Aristotelian model (form informing matter) for understanding Montaigne's thought.

In the essay 'De l'oisiveté' (I. 8), Montaigne famously characterizes his own mind in the state of idleness using the image of weeds that proliferate in a fallow land, and calls for the need to 'subject' and 'subdue' his own idle thoughts in some process of intellectual cultivation analogous to agriculture. 'Comme nous voyons des terres oysives, si elles sont grasses et fertilles, foisonner en cent mille sortes d'herbes sauvages et inutiles, et que, pour les tenir en office, il les faut assubjectir et employer à certaines semences, pour nostre service' ('Just as fallow lands, when rich and fertile, are seen to abound in hundreds and thousands of different kinds of useless weeds so that, if we would make them do their duty, we must subdue them and keep them busy with seeds specifically sown for our service').[23] If agriculture consists in reigning in of the proliferation of weeds and subjugating them through cultivation, then the third element of the simile—I pass over the second part, pregnant women—the mind ('des espris'), is reined in by the activity of writing, but these acts of 'reining in' extend rather than eliminate the proliferation, and benefit from having been infused by an atelic vegetal form. In 'De l'experience' (III. 13), ideas themselves take on vegetal shape. Montaigne adapts the notion of plant grafting to describe how opinions proliferate: 'Nos opinions s'entent les unes sur les autres. La premiere sert de tige à la seconde, la seconde à la tierce' ('Our opinions graft themselves on to each other. The first serves as stock for the second, the second for the third').[24] Montaigne also uses the image of grafting and the hybridity of grafted plants to describe the process of writing, in which ideas are similarly 'grafted' onto those borrowed from other sources. Writing benefits from grafting and from the proliferation of images in the mind. *The Essays* are thus a mental exercise in which the strangeness of the mind is domesticated (but not eliminated), just as weeds are in agriculture.[25]

More recently, scholars have homed in on one of the sources mentioned by Jeanneret as a source of the author's reflections on matter in flux, namely Lucretius's poem *De rerum natura*. Montaigne read Lucretius's *De rerum natura*, re-discovered in Europe in 1417, in Denis Lambin's erudite 1563 Latin

22 On this formulation of Aristotelian relation of matter and form, see Jeanneret, *Perpetual Motion*, p. 97.

23 Montaigne, *Les Essais*, 32a / *The Complete Essays*, p. 31.

24 Montaigne, *Les Essais*, 1069c / *The Complete Essays*, p. 1212.

25 For a similar reading of the weeds image in 'De l'oisiveté', see Regosin, *Montaigne's Unruly Brood*, p. 156.

edition (his personal annotated copy of Lambin's Lucretius was discovered in 1989 and published by Michael Screech in 1998).[26] In the wake of this discovery, scholars have revised the earlier consensus that Montaigne had read the book in verse, mostly for the pleasure of its poetic style, and rejected the teachings of Epicurean philosophy.[27] Montaigne explicitly dismisses the theory of atoms a number of times, in particular asking why, if the movement of atoms is supposed to create everything, we do not see these elements engender something like a house or pair of slippers:[28] 'Si les atomes ont, par sort, formé tant de sortes de figures, pour quoy ne se sont ils jamais rencontrez à faire une maison, un soulier? Pour quoy, de mesme, ne croid on qu'un nombre infini de lettres grecques versées emmy la place, seroyent pour arriver à la contexture de l'Iliade?' ('if atoms do, by chance, happen to combine themselves into so many shapes, why have they never combined together to form a house or a slipper?').[29] This question indeed shows Montaigne's doubt about the efficacy of atoms and their ability to form things—in particular human artifacts—through their contingent movement (he explains Epicurean mechanics, especially the Greek notion of *clinamen*, or Latin *inclination*, in the previous sentence); nonetheless it also reserves the possibility that natural things, unlike products of human art, come about as a result of movement of physical particles.

George Hoffmann has argued that, while Montaigne retained a sceptical reserve toward many of its doctrines, Epicurean philosophy allowed him to develop his particular kind of naturalism. According to Hoffmann, Montaigne embraced the explanation of physical phenomena through natural rather than divine or first causes, while also eschewing proto-empiricist reasoning, and he drew on the theory of atoms to argue that, ultimately, natural causes could not be known.[30] Montaigne used atomic movement when it was convenient to undermine proto-empiricist arguments (for example, by medical doctors) about causes, in agreement with Galenic suspicion toward empiricism. He shows that Montaigne, notably in 'De l'experience' (III) turns the movement of everything into a tool of consolation in times of instability,

26 Screech, *Montaigne's annotated copy of Lucretius*.

27 See especially Hendrick, 'Montaigne, Lucretius and Skepticism', pp. 139–152 and Moore, 'Lucretius and Montaigne', pp. 109–114. Michael Screech, although confirms that Montaigne was an attentive and informed reader of Lucretius, still sides with this side of the argument in his Introduction to *Montaigne's annotated copy of Lucretius* (pp. 44–45).

28 See, for example, V. 511 and V. 544–545.

29 Montaigne, *Les Essais*, 544–545 / *The Complete Essays*, p. 612.

30 Hoffmann, 'The Investigation of Nature…' in *The Cambridge Companion to Montaigne*, pp. 163–182.

but argues that Montaigne most fully embraces the materialist doctrine of chance governing the physical world when it comes to the workings of the human mind, including his own. How then does Montaigne's naturalism and his scepticism toward causal determinism elevate the humble plant, with its sap and rootedness in the soil, to the status of exemplary figure of animation in *The Essays*?

Montaigne's description of the tree in the 'Apologie', which forms the second part of a tri-partite image borrowed from Sextus Empiricus, supports Hoffmann's claim that 'Montaigne's thinking about randomness generally follows Epicurean arguments instead of the sceptical modes from Sextus Empiricus that one might have expected'.[31] Montaigne writes, translating and paraphrasing verses from the *Hypotyposes*:

> comme nous voyons du pain que nous mangeons: ce n'est que pain, mais nostre usage en faict des os, du sang, de la chair, des poils et des ongles:
> *Ut cibus, in membra atque artus cum diditur omnes,*
> *Disperit, atque aliam naturam sufficit ex se.*
> L'humeur que succe la racine d'un arbre, elle se fait tronc, feuille et fruit; et l'air n'estant qu'un, il se faict, par l'appliquation à une trompette, divers en mille sortes de sons: sont-ce, dis-je, nos sens qui façonnent de mesme de diverses qualitez ces sujects, ou s'ils les ont telles?
>
> (Rather like bread when we eat it; it is one thing, bread, but we turn it into several: bones, blood, flesh, hair and nails.
> *Ut cibus, in membra atque artus cum diditur omnes,*
> *disperit, atque aliam naturam sufficit ex se.*
> (Like food, which spreads to all limbs and joints, destroys itself and produces another substance.)
> Moisture is sucked up by the roots of a tree: it becomes trunk, leaf and fruit; air is one, but when applied to a trumpet it is diversified into a thousand kinds of sound: it is our senses (I say) which similarly fashion such objects with diverse qualities or do they really have such qualities?)[32]

Is it our senses that fashion matter into diverse qualities, or is this diversity inherent to matter? Montaigne brings his examples about the unreliability of knowledge to a tantalizing and counterintuitive conclusion. He ends with a question that casts doubt on the main (Pyrrhonic) lesson of the essay.

31 Ibid., p. 173.

32 Montaigne, *Les Essais*, p. 599ab / *The Complete Essays*, p. 677.

This tripartite image, which draws on a passage from Sextus Empiricus's *Hypotyposes*, follows a lengthy discussion of the unreliability of the senses, the most powerful argument in the arsenal of ancient Sceptics for questioning the validity of dogmatic knowledge.[33] In this text, the argument about the diversity of perception (according to which the perceived quality of a thing depends on the different organs or the make-up of different animals), gives way to an argument about the diversifying power of physiological and physical processes, such as the ingestion and digestion of food in human bodies, nutrition in plants, and the movement of air in a trumpet. We might wonder if Montaigne was aware of its ambiguity.

All three examples here take us from an argument about the unreliability of the senses to a different one about the flux of matter and the constant metamorphosis of things in the physical world. Montaigne juxtaposes examples in which some substantial and familiar material thing transforms into some seemingly distinct thing: bread morphs into bones, blood, flesh, hair, and nails in the body; the sap metamorphoses into trunk, leaves, and fruit in the tree; and air becomes the different sounds of a melody in a trumpet. This passage is imbued with a rich poetic and material suggestiveness that questions which comes first: the embodied being that takes in nutrition—like the human being or the tree—or the very process of becoming itself? The passage transmits a positive fascination with becoming, to the point that the wondrous quality of the transformation of the bread into human's or animal's body parts (depending on how we read 'nostre') rivals transubstantiation. The last example—air passing through the trumpet—is especially telling: what we are admiring, the music, is a vibration of air channelled through the trumpet (after it has been channelled through the lungs and the human body), rather than the trumpet itself (an product of human culture and art) or indeed the human body. It is also remarkable that the first two images linked to the Aristotelian nutritive or *vegetable* soul are given a decidedly Epicurean spin here, especially as the various transformations of matter produce mostly pleasure and wonder.[34]

Plants, as they appear in *The Essays*, correspond to the author's interpretation of the Epicurean world of atoms: he adapts them to a natural

33 'That it has only one quality can be argued from what we said before about the nourishment dispersed in our bodies and the water dispersed in trees and the breath in flutes and pipes and similar instruments'. See *Outlines of Scepticism*, p. 26.

34 In fact, Screech suggests that this passage also owes much to several verses of Lucretius. Lucretius, *On the Nature of Things / De rerum natura*, p. 111. This passage compares the absorption of food and its transformation into members of the body (a kind of death itself) to the mortal soul 'seeping' into the different parts of the body.

world that is familiar (as its objects are present in the countryside, on the fields, in the garden), but which also becomes defamiliarized as matter in movement. Montaigne's botanical imagery is defined by this Epicurean naturalization of the Aristotelian notion of the nutritive soul, rather than being strictly naturalist in a proto-empiricist sense, in which matter was seen to move in the context of causal relations. Montaigne's plants are part of the natural world; their 'movements'—the formation of leaves and fruit, for example—are not shown as products of some overarching art or reason. Thus Montaigne anticipates the naturalism of the seventeenth-century botanist and founder of the Jardin des plantes, Guy de la Brosse (1586–1641), who calls them stars on Earth in his Preface to *De la nature, virtu et utilité des plantes* (1628). While not an Epicurean, but inspired by alchemy and the practice of observing nature, De la Brosse attributes to plants an immaterial soul, but this soul is unique to the individual plant, a source of a fascinating singular power in each plant. For Montaigne, plants participate in the cosmic movement Lucretius undertook to describe after Epicurus, while also bringing it down to a scale that is terrestrial and human. They are in constant movement that is slow enough to appear stable. Plants thus figure the movement of matter, the sudden changes of the human psychology, and the life of generations. Montaigne approaches life and the living—sickness, the mind, relation to offspring, social life, and, especially, as we shall see, the relation between fathers and children—with a materialist interest in random movement borrowed from Epicureanism, the same mobility that he accords to the natural world, all the while eschewing the proto-empiricist attitude of attributing (based on observation) causes to these movements of matter.

Recently, Rafal Krazek has argued that Montaigne's thought has been deeply informed by Lucretian thought, devoting a whole chapter to Montaigne's naturalism, which he sees manifesting itself in the Epicurean recognition that nature comprises culture, civilization and the individual as well.[35] Plants thus help bring Montaigne's mind back to the earth and prevent it from seeking the 'infinite', an inaccessible transcendence that Krazek identifies as the fallacy of thought in a naturalist, Epicurean system. Yet while Krazek emphasizes the Epicurean directive that thought needs to return to the body,[36] long recognized as characteristic of Montaigne's mode

35 Rafal Krazek, *Montaigne et la philosophie du plaisir*, esp. pp. 118–119. He also acknowledges Montaigne's syncretism and the importance of Pyrrhonic doctrines, p. 128, and offers a longer review of the history of interpreting Montaigne's indebtedness to Lucretius's presentation of Epicureanism. Ibid., pp. 35–51.

36 Krazek, *Montaigne et la philosophie du plaisir*, p. 122.

of inquiry, he, too, misses Montaigne's innovative idea that plants—less than embodied in the animal sense, yet comprised of the morphology of a body—can also serve as figures of thought that, while not exactly inspired by Epicurean figures or teachings, can be considered Epicurean. Hoffmann emphasizes how 'Montaigne's Epicurean naturalism applied itself not so much to nature as to the nature of humans, not so much to the physical world as to the mental world'.[37] Unlike Krazek and Hoffman, I do not seek a reading that locates or reinscribes specific Epicurean moral teachings in the *Essays*, nor even an extension of atomism into Montaigne's social world. I argue that plants not only take Montaigne back to his body but, in their ability to figure matter in movement, indicate a place *outside* culture and art that, in turn, impinges on human culture and on the body's *habitus*, pleasures, and tastes. This implies that the famous self in movement described by Starobinski can now also be seen as part of a larger movement, which does not derive from the unstable institutions and upheavals of the social political world, nor from the processes of the human body (aging, illness, etc.), but is manifest in the material world that Montaigne sees as (by and large) untouched by human art—and which is best exemplified by plants.

Although Montaigne's plants, in their constant movement, are Epicurean, it is important to note that they correspond to the depiction of matter made up of atoms, *not* to the depiction of *plants* in *De rerum natura*. Called 'the philosophy of the garden', since Epicurus conducted his teaching in the groves of his private garden outside the city of Athens, plants in Epicureanism are predominantly associated with bucolic and agricultural landscapes. This is the case in the passage that, according to Hoffmann, may have inspired Montaigne to think as a non-empiricist naturalist:

> And so, we all arise from sky-born seed.
> There is one father for all. When the fostering Earth,
> Our mother, takes within her his moist droplets,
> Grown big, she bears the glossy corn and the orchards,
> The human race and every kind of beast,
> Proffering food for all to feed their bodies
> And live sweet life and branch out into offspring. (II. 991–997)[38]

Lucretius's agricultural image of the genesis of living beings out of the joining of atoms of the soil and 'seeds' falling from the sky takes advantage

37 Hoffmann, 'The Investigation of Nature', p. 173.

38 Lucretius, *On the Nature of Things / De rerum natura*, p. 85.

of the vegetal connotations of the word *semina* (Lat. 'seeds', 'sperm', 'elements', 'shoot' or 'graft' and 'race'), which is also one of his terms for atoms (e.g. *semina rerum*). This image of 'sky-born' seeds falling into the earth is reminiscent of agricultural cultivation, and the passage continues to describe all generation according to this analogy: glossy corn, orchards, beasts, and the human race alike are all born this way, borne by 'mother' earth and 'fathered' by heavenly seeds. A second order of agricultural production makes sure that all these beings are fed. Thus in Lucretius, an agricultural metaphor ensures that not only are new things are constantly generated but also that there is a relative orderliness and consistency to this process (the world does not change). Although every creature is bound to die and either 'turn again to earth' or 'be| Brought back into the temples of the sky' (II, 999–1001), there will be lives lived and food to sustain them.[39]

While in Lucretius the *semina* are images of atoms setting in motion an agricultural event, Montaigne takes these images of 'semence' and makes them a lot more concrete in his vegetal figures. His plants are not visible analogies for the workings of invisible (and thus abstract) atoms, but physical bodies made up of sap or Galenic humours. Nor do they reliably provide us with agricultural products, let alone engender the world in its reassuring stability, but rather gesture toward a world that is shifting, transforming, and changing. Hans Blumenberg concentrates on such Lucretian images of a world of bucolic stability, and identifies the arrival of modernity with the loss of this 'therapeutic' belief in a reliable and predictable nature, generating a strong desire for the technological mastery of nature. In *Enchantment of Modern Life,* meanwhile, Bennett proposes an alternate reading of Epicurean atomism as a contingent, unpredictable process and offers a view of modernity that is not defined solely or predominantly by the instrumentalization of matter and nature.[40] Montaigne shows no interest in the bucolic images of agriculture in *de rerum natura*—in fact he defamiliarizes agriculture to which he at first relates, through his ignorance—but endows his plants with the kind of unpredictable and often satisfying movement that Bennett argues atoms possess in Epicurean thought. Montaigne's 'semence' is vigorously animate matter that ushers us into a modernity as envisioned by Bennett,

39 This is what Blumenberg describes as the therapeutic character of Epicurean philosophy, which, although it rejects providence, still reassures its practitioners that the chance movement of atoms provides that there will always be human beings and everything that they need. While Blumenberg claims that this therapeutic function characterizes all of Epicurean philosophy, I underscore only that plants in *De rerum natura* represent such stable, bucolic or agricultural spaces. Blumenberg, *The Legitimacy of the Modern Age,* p. 165.

40 Bennett, *Enchantment of Modern Life,* pp. 72–75.

rather than the Cartesian one favoured by Blumenberg. Together with de la Brosse, Montaigne accords agency to plants rather than taking the route of the forming proto-empiricist botany that avails itself of plants as objects to observe, collect, describe, and gather knowledge about. Montaigne's plants intervene in his writing. As we have seen in the oft-cited image of the mind as fallow land overgrown with weeds, and also in images that refer to the vegetal structure of the book or the grafts of our opinions, vegetal images particularly abound in discussions of the human mind. It is these grafts and weeds that the author engages with; writing is not simply the agency of a rational form upon them, but is comprised of both the vegetal profusion of images and words—which resemble vegetal matter in its atelic shape and process—*and* the human art and *habitus* of giving them a second, more culturally recognizable, domesticated shape.[41] Describing his lively mind in old age, Montaigne returns to the vegetal image in 'Sur des vers de Virgile' (III): 'Puisque c'est le privilege de l'esprit de se r'avoir de la vieillesse, je luy conseille, autant que je puis, de le faire: qu'il verdisse, qu'il fleurisse ce pendant, s'il peut, comme le guy sur un arbre mort' ('let it meanwhile sprout green and flourish, if it can, like mistletoe on a dead tree').[42] The image of the thriving human mind in an aging body, like a green clump of mistletoe on a dead tree, returns us once again to the vegetal alive-ness and mobility of the mind (all of it in the context of reading erotic poetry and without any suggestion of an immaterial rational mind).[43] Indeed, vegetal images describe the mind in a number of other contexts, including the relationship of children and fathers and one's relationship with oneself. Humoral trees prefigure the transmission, from generation to generation, of not only psychological traits but also a certain patrilineal organization of society and politics. Montaigne, of course, did not think of the biological process of genetic inheritance, but rather of a complex process of imitation, upbringing, and transmission of blood, which in the Galenic system was conceived of as the humour that carried family traits (and family honour). In 'De l'affections des peres aux enfants' (II. 8), Montaigne calls authoritative, austere fathers 'scarecrows in a field of flax' ('vrais espouvantails de

41 My reading here confirms Brancher's argument that 'Il [the vegetal] se prête donc particulièrement bien à la désignation métaphorique de genres qui relèvent de l'accumulation composite et de la collection protéiforme, où le retranchement d'une partie n'affecte pas la vitalité de l'ensemble', (The vegetal realm thus finds a great expression in the metaphorical designation of genres that fall under composite accumulation and protean collection, where the removal of one part does not affect the vitality of the whole, my translation), p. 18.

42 Montaigne, *Les Essais*, 844b / *The Complete Essays*, p. 951.

43 See Conley's reading of this image, 'Montaigne mocqueur', p. 96.

cheneviere'), suggesting that their authority is soon to be outgrown by their children's sheer physical and mental development.[44]

Montaigne's vegetal psychology is most explicit in 'De l'art de conferer' (III. 8), where he makes the striking claim that he is capable of looking at himself as a tree (in response to the accusation of self-love he has received as the author of *The Essays*): 'Je ne m'ayme pas si indiscretement et ne suis si attaché et meslé à moy que je ne me puisse distinguer et considerer à quartier: comme un voisin, comme un arbre' ('I do not love myself with such a lack of discretion, nor am I so bound and involved with myself, that I am unable to see myself apart and to consider myself separately as I would a neighbour or a tree').[45] It is also in this statement that the transformation of psychology into an ethics becomes most clear. Montaigne does not settle on a Neoplatonist image of the tree as an upside-down human being, a reversal of human dignity (like Pico della Mirandola and other Renaissance Neoplatonists did). For Montaigne's tree is neither a reversal nor a privative repetition of the human: it both prefigures and resembles what we consider human culture, and stands at a distance from it; it differs from it like the movement of atoms differs from the shoes and the poems that only human art can make. The self can thus be both similar to and differ from a tree, a distance that is analogous to an ethical and political relationality implied by the first term of the simile, a neighbour (whose person also connotes relative closeness but also difference—for example, between interest and religion). This is how Montaigne reiterates his injunction in 'De la cruauté' to respect plants and trees, suggesting that when we look at plants, we both find an intimate similarity and a difference. Thus a tree is not simply something familiar and close by in the phenomenological sense, but is also at a distance, like other people are whose minds we cannot read or who have different customs or faiths (as Montaigne often reminds us), because the material world remains essentially unfamiliar and unlike human artefacts. The tree also implies stability in motion because, as Hoffmann aptly puts it, in an Epicurean physical world, stability is 'very slow movement', and this slow movement of the tree (as sap) provides a central axis and a verticality within movement itself. But this slow movement is not only a comforting metaphor for a society in crisis; it also involves the self in something that is different from its social conception, literally transfroming it into the

44 Montaigne, *Les Essais*, 393a, p. 441. Montaigne carefully clarifies throughout the essay that he considers children as rational beings, even to the point of confessing his utter disinterest in new-born infants: 387a, p. 435.

45 Montaigne, *Les Essais*, 942c / *The Complete Essays*, p. 1067.

movement that is a tree. For a tree is movement around a vertical line, a centered, stable movement that Starobinski attributes to the 'self' as represented in *The Essays*. Is this self, then, vegetal?

Reading Montaigne after Ecology

The distance that separates the movement of matter in nature from human society, culture, and art allows the author to turn away from the latter by turning to the former; however, if we frame our view with the concepts of an ecological thought—the anachronistic gesture that this volume invites us to make—then this plant turn is in no way devoid of ethical implications. For us, after ecology, Montaigne represents a voice in early modernity that speaks about our relation to a nonhuman and non-social world, conceived of as undefined by our mastery. Montaigne's plants in movement recall Bennett's turn of phrase 'vibrant matter', her playfully strategic conceptual response to the modern, disenchanted, mechanistic view of the world. Bennett rejects the notion of nature as external to us and externalized by modernity, and she does so in order to challenge our received notions of the capacity to act on and within the world as a uniquely human prerogative. Her thinking is geared towards affective interactions. For Bennett, the fascination with things—and a recognition of their power—can and should become part of a more ecological way of being in the world. She writes, 'This sense of a strange and incomplete commonality with the outside may induce vital materialists to treat nonhumans—animals, plants, earth, even artefacts and commodities—more carefully, more strategically, more ecologically'.[46] 'Vibrant' matter is animated, lively, and does not, moreover, lend itself to objectivation and instrumentalization. Although separated by centuries, and although there are too many differences in thought to list, Montaigne shares with the vital materialist Bennett the commitment to acknowledging our intimate relationship with things not generally or easily considered as ethical others. They also both reject linear history and offer us passionate, attentive, and unique readings of Lucretius. For Montaigne, as for Bennett, this stems from philosophical, ethical, and political, as well as aesthetic attunement to the material world that, in Bennett's case, brackets scientific empiricism (which Montaigne, at the other end of the historical development, does not yet embrace). More specifically, the resemblance between the tree and the human being that Montaigne identifies is based on the sympathy of

46 Bennett, *Vibrant Matter: A Political Ecology of Things*, p. 18.

shape or arrangement that Bennett highlights as a defining trait of vibrant matter: 'Sensuous experience is central to enchantment, but, of course, not all sensuous experience enchants. Enchantment seems to require, among other things, the presence of a pattern or recognizable ensembling of sounds, smells, tastes, forms, colours, textures'.[47] Plants, one could argue, serve the function of presenting precisely such patterns, where movement produces shape, and shape coheres into a morphology.

Although none of the famous exempla of the 'Apologie' concern plants, plants are nonetheless included in the logic of both resemblance and ethical distance, which is expressed in much more explicit terms in the case of animals. 'Nature clasps all her creatures in a universal embrace; there is not one of them which she has not plainly furnished with all means necessary to the conservation of its being' ('Nature a embrassé universellement toutes ses creatures; et n'en est aucune qu'elle n'ait bien plainement fourny de tous moyens necessaires à la conservation de son estre').[48] Citing the received view that clothes are necessarily provided by human art because, while '[n] ature has clad all others with shells, pods, husks, hair, wool, spikes, hide, down, feathers, scales, fleece or silk' ('de coquilles, de gousses, d'escorse, de poil, de laine, de pointes, de cuir, de bourre, de plume, d'escaille, de toison et de soye'), the naked human body lacks protection, and arguing that the naked human skin is not without the endurance provided by these other (vegetal and animal) shells, Montaigne ends the debate with a conclusion drawn from Lucretius: 'For every creature feels the powers at its disposal' ('Sentit enim vim quisque suam quam possit abuti').[49] This argument culminates in the reassertion, in concert with Lucretius (this time Book 5 of *De rerum natura*), of the power of the Earth, helping Montaigne to discredit human arts, from agriculture to the custom of making and wearing clothes.[50] Montaigne claims here (and elsewhere devotes a whole essay to this claim) that human beings are properly protected by their skin, and this list includes items that serve as human clothing but point to animal or vegetal production. As Hoffmann argues, Montaigne's aim is not to assert, as Aristotle does, that a human *technê* (art or craft) imitates a certain rationality that belongs to nature. Instead, Montaigne discovers a 'force' or 'power' in nature that is different from the Aristotelian *technê* and is produced by the movement of matter. Animal skins and vegetal husks are

47 Bennett, *Enchantment*, 36.
48 Montaigne, *Les Essais*, 509 / p. 456a.
49 Montaigne, *Les Essais*, 511.
50 Ibid.

thus not inert matter but visible manifestations of this active power. Our skins are also such instances of powerful matter. While many have noted Montaigne's interest in discrediting the superiority of human arts, this attribution of a positive force to natural things has only been explored in his examples of animals. When taken outside a theological frame, as those who highlight Montaigne's naturalist thinking suggest, the vegetal pods, animal hides, and, similarly, the human skin are all produced through such movements, whose causes remain ambiguous while the effects remain observable. Thus, even though Montaigne abandons references to plants when he argues that animals have capacities of reasoning and moral and social sentiments, plants serve him as the perfect example of the Epicurean movement of atoms producing things that have lives of their own, whether that be a tree, the human mind, or our skin. We are endowed with a husk called skin that is alive, part of the morphology of the human, as vegetal husks and animal hide define plants and animals.

Although Montaigne uses vegetal imagery to turn his gaze, in a presumptuous self-absorption, towards himself, his 'portrait', his own book, his father, and the patrilineal family,[51] my analysis has aimed to show that Montaigne is concerned about a material world that is distinct from our perceived social and cultural world and is aware that even slow movement can produce surprising changes in those seemingly stable familial and social structures (e.g. a feud in the family from generational tensions or strange and disruptive ideas from the constant grafting of opinions). Likewise, it would be a mistake to read the vegetal merely as an image of the random movements to which the human mind is subjected. Not only are Montaigne's images often very concrete, based on a kind of perceptive interest in plants (one that is also attested by his recording of his conversation with the Pisan woodworker), but they also serve as helpful tools for locating oneself on the ground, on the earth. Thus plants *do* resemble human beings in their location on earth, their vertical position, and their constant multi-directional movement. The soil, to which plants are so intimately tied, should not be conceived as inert matter (*pace* Aristotle), but a place of rich production and unpredictability. Montaigne's frequent fascination with the naked skin ties in with both the foregrounding

51 See Regosin's analysis of Montaigne's claim that his book is his progeny. Regosin points to Ovid's Myrrha, who turns into a tree (and gives birth, as tree, to Adonis) after her incestuous union with her father. Regosin cites this myth as the image of self-love that haunts Montaigne's reflections on progeny and his book; one could also cite the seventeenth-century libertine author Cyrano de Bergerac's *Voyage au soleil*, in which trees (oak trees this time) harbour incestuous and otherwise transgressive desires and engender a neo-Ovidian world of libidinal difference, a tradition to which Montaigne serves as a bridge. Regosin, *Montaigne's Unruly Brood*, pp. 133–136.

of this human husk as vibrant matter and his rejection of the technological model that reduces this matter into something to be manipulated through human art. Our ontological proximity and morphological similarity to plants renders them similar, while their participation in a cosmic movement of matter renders us unfamiliar—both of which realizations inspire Montaigne's unique ethics. The self of the *Essays* is vegetal movement.

Works Cited

Bennett, Jane, *The Enchantment of Modern Life: Attachments, Crossings, and Ethics* (Princeton, N.J.: Princeton University Press, 2001).

———, *Vibrant Matter: A Political Ecology of Things* (Durham, N.C.: Duke University Press, 2010).

Blumenberg, Hans, *The Legitimacy of the Modern Age* (Boston, MIT Press, 1983).

Brancher, Dominique, *Quand l'esprit vient aux plantes: botanique sensible et subversion libretine (XVII-XVIIIe siècles)*, (Geneva: Droz, 2015).

Conley, Tom, 'Civil War and French Better Homes and Gardens', *The South Atlantic Quarterly* 98.4 (Fall 1999), 725–759.

Desan, Philippe, *Montaigne: A Life* (Princeton: Princeton University Press, 2017).

———, ed., *Dictionnaire de Michel de Montaigne* (Paris: Garnier, 2016 (2007)).

Grew, Nehemiah, *Anatomy of plants*, 1682.

Hendrick, P.J., 'Montaigne, Lucretius and Skepticism: And Interpretation of the "Apologie de Raimond Sebond"', *Proceedings of the Royal Irish Academy: Archeology, Culture, History, Literature* 79 (1979), 139–152.

Jeanneret, Michel, *Perpetual Motion: Transforming Shapes in the Renaissance from da Vinci to Montaigne* (Baltimore: Johns Hopkins University Press, 2001).

———, *Perpetuum mobile: Métamorphoses des corps et des oeuvres de Vinci à Montaigne* (Paris: Éditions Macula, 1997).

Krazek, Rafal, *Montaigne et la philosophie du plaisir: Pour une lecture épicurienne des* Essais (Paris: Garnier, 2011).

Langer, Ullrich ed., *Cambridge Companion to Montaigne* (Cambridge: Cambridge University Press, 2005).

Leonardo da Vinci, *Trattato della Pittura tratto da un codice della Biblioteca Vaticana e dedicato alla maestà di Luigi XVIII re di Francia e di Navarra- Disegni che illustrano l'opera del trattato della pittura tratti fedelmente dagli originali del Vodice Vaticano* (Roma: De Romanis, 1817).

L'homme plante (Potsdam: Voss, n.d.).

Long, Kathleen Perry, ed., *High Anxiety: Masculinity in Crisis in Early Modern France* (Kirksville, MO: Truman State University Press, 2002)

Lucretius, *On the Nature of Things / De rerum natura*, trans. by Anthony M. Esolen (Baltimore: The Johns Hopkins University Press, 1995).
MacPhail, Eric, 'Montaigne's New Epicureanism', *Montaigne Studies*, 12.1–2 (2000), 91–103.
Marder, Michael, *Plant-Thinking: A Philosophy of Vegetal Life* (New York: Columbia University Press, 2013).
Michel de Montaigne, *The Complete Essays*, trans. by Michael A. Screech (London: Penguin, 1991; revised 2003).
———, *Journal de Voyage*, ed. by François Rigolot (Paris: PUF, 1992).
———, *Les Essais. Livres I–III*, ed. by Pierre Villey (Paris: PUF, 2004).
Moore, W.G., 'Lucretius and Montaigne', *Yale French Studies* 38 (1967), 109–114.
Myers, Natasha, 'Photosynthetic Mattering: Rooting into the Planthroposcene', in *Moving Plants*, ed. by Line Marie Thorsten (Naevsted: Roennebaeksholm, 2009), pp. 123–127.
Nealon, Jeffrey T., *Plant Theory: Biopower and Vegetable Life* (Stanford: Stanford University Press, 2015).
Armaingraud, A., ed., *Oeuvres complètes de Michel de Montaigne* (Paris: Conrad, 1932).
Regosin, Richard L., *Montaigne's Unruly Brood: Textual Engendering and the Challenge to Paternal Authority,* (Berkeley: University of California Press, 1996).
Screech, Michael A., *Montaigne's annotated copy of Lucretius: A transcription and study of the manuscript, notes and pen-marks*, Travaux d'humanisme et Renaissance (Geneva: Droz, 1998).
Sextus (Empiricus), *Sextus Empiricus: Outlines of Scepticism*, Cambridge Texts in the History of Philosophy, ed. by Julia Annas and Jonathan Barnes (Cambridge: Cambridge University Press, 2000).
Shannon, Laurie, 'La chatte de Montaigne', in *Dictionnaire de Michel de Montaigne*, ed. by Philippe Desan, (Paris: Champion, 2004), p. 162.
Starobinski, Jean, *Montaigne en movement* (Paris: Gallimard, 1993 revised edition; first published in 1982).

About the Author

Antónia Szabari (Associate Professor of French and Comparative Literature, University of Southern California) works on early modern French literature and, in a broader fashion, on the culture of early modern Europe. *Less Rightly Said: Readers and Scandals in Sixteenth-Century France*, her study of French religious and political satire, was published by Stanford University Press in 2009. It shows how pamphlets helped to transform the theological and dogmatic function of religion in the sixteenth century into a way of constructing, seeing, and inhabiting a world of differences. Together with Natania Meeker, she is currently at work on book tentatively titled *The Animated Plant: Vegetal Imaginaries from Early to Late Modernity*, which examines the plant as the object of early modern speculations about the nature of matter and the material basis of life, the 'animation' of plants in early modern science fiction and modern time lapse photography, and why the plant has become the favorite trope of ecological thought in art and philosophy today.

Epilogue

Louisa Mackenzie

I admit to feeling a little embarrassed—though flattered, of course—at finding myself face to face with a rhetorical flourish I came up with to conclude a Sunday morning MLA panel. 'What can early modern French literature do for ecocriticism?' It is easy to ask sweeping questions like this one—*verba volant*, after all. It is much harder to respond as this volume's editors did initially, and their contributors subsequently, with serious scholarly engagement, including troubling (in a very welcome way) some of the assumptions of the question itself. The result is this innovative and diverse collection that does the work my question merely gestured at before retreating. The authors have all done the hard work; it is a pleasure to write this epilogue and I hope I can do some justice to their insights.

Renaissance French humanism is of course a distant relative of our own intellectual field, the humanities 'with an accent' as Phillip and Pauline put it. And this book shows, I think, that humanist thinking then and now is always already é/ecological. As I tried to pull together the many connections made here between past and present, I realized that the authors were also making connections between humanistic and ecological thinking as modes of relating and being in the world. Wondering where to start with this notion, I found myself thinking, perhaps predictably, of Joachim Du Bellay. Here are the quatrains of the sonnet, number 38 from *Les Regrets,* which sprung to mind:

> Ô qu'heureux est celui qui peut passer son âge
> Entre pareils à soi! et qui sans fiction,
> Sans crainte, sans envie et sans ambition,
> Règne paisiblement en son pauvre ménage!
> Le misérable soin d'acquérir davantage
> Ne tyrannise point sa libre affection,
> Et son plus grand désir, désir sans passion,
> Ne s'étend plus avant que son propre héritage.[1]

1 Du Bellay, *Les Regrets et autres œuvres poétiques*, ed. by Screech and Jolliffe, p. 38. My translation.

Goul, P. and P.J. Usher (eds.), *Early Modern Écologies. Beyond English Ecocriticism.*
Taylor & Francis Group, 2020
DOI 10.5117/9789462985971_EPI

(Oh happy is he who can spend his life | With his peers! And who without lies | Fear, desire or ambition | Rules peacefully in his modest household! The wretched business of acquiring more | Does not dominate his free affection | And his greatest wish, devoid of passions | Does not extend further than his own inheritance).

I believe this to be a profoundly ecological poem, as well as a humanist one. It is a *Beatus ille* that opposes acquisitive thinking, structurally and thematically, to a fantasy of modest self-sufficiency. While Du Bellay's moral defence of autarky seems to be far removed from what modern readers might think of as ecological (there is no nonhuman landscape or animal in this sonnet) it is a profoundly early modern French ecological poem. Du Bellay's hypothetical *beatus* does not desire more than what his family has given him, he does not have to travel abroad or depend on others, he is content to stay at home with people like him and simply subsist. (Readers who appreciate the relation between metrics and message will note the rhyme positions of the key words *davantage, ménage,* and, later, *héritage,* the metrics mapping out the poem's moral economy and ecology in ways to which critics such as Tom Conley have trained us to be attentive). The poet's articulation of sustainable household management is an excellent example of the ideal early modern *oikos.* This, the Greek word meaning household or place to live, is of course the root of the morpheme *eco* in words such as economy and ecology. These two eco-words, as Conley reminds us in his chapter here, are conceptually related. To be ecological is to be economical; we might call it an ethics of sustainability, a modest relation to habitat subtracted from the logics of (proto)capitalism.

These essays variously show that ecology is in fact a much more appropriate concept for early modern France than is environment. Vin Nardizzi has recently cautioned early modernists to 'unlearn'—with Wendell Berry and Michel Serres—the semantic field of environs, inasmuch as it posits humans in a centre and other objects outside.[2] This book, I think, encourages us to slow down with Renaissance French texts and unlearn their *environs* while tracing their *oikos,* their ecologies and economies. Tom Conley's chapter on Olivier de Serres is key here: not only is Serres's work subtitled with the word *ménage*, it is an entirely agrilogistic text, which sees human use of natural resources—economy—as the only thinkable relation to land—ecology—and which inserts itself *into* that ecology. Nature is not

2 Nardizzi, 'Remembering Premodern Environs', pp. 179–183.

around (environ) humans, it is coterminous and coproductive *with* them. Prepositions are more important that they might seem.

In a section of her chapter that particularly struck me, Antónia Szabari looks carefully at the preposition *avec* in a section of Montaigne's 'Apology for Raymond Sebond'. The Apology is often read these days as a catalogue of human–animal intimacies; by slowing down with one small word, Szabari invites us to think about the with-ness of human–*plant* relations. What happens when a relation that seems merely prepositional becomes instead a vibrant, dynamic entanglement? It becomes an invitation to ethics. To think ecologically in early modern France is to think through an ethos of life itself, about how humans inhabit, manage, and relate to, their dwelling places: how they live *with* and *in*. As Joseph Campana and Scott Maisano put it in the introduction to their important collection *Renaissance Posthumanism,* reading Renaissance humanist texts on their own terms can reveal 'ideas of "the human" as at once embedded and embodied *in*, evolving *with*, and decentered *amid* a weird tangle of animals, environments, and vital materiality'.[3]

Not surprisingly, our authors here think frequently with Bruno Latour, Timothy Morton, and Jane Bennett; with naturecultures, networks, actants, the modern constitution, dark ecology, agrilogistics, hyperobjects, vibrant materialism. This book reconfirms what I suspect most early modernists sometimes feel when confronted with an exciting new theoretical frame: it's not that new, and Renaissance humanists already knew it. Of course, this rather grumpy reaction is not particularly useful: it doesn't get us beyond a critical habit that simply backdates contemporary theory. The chapters here do much more than that. They demonstrate that primary engagement with Renaissance humanism within the scholarly frames of ecocriticism, Animal or Plant Studies, posthumanism, etc., complicate the sometimes simplistic characterizations of its legacy in more than one influential contemporary theoretical work.[4] Renaissance humanism cannot be used as a foil against which a new posthumanism arises; it is not a stand-in for human exceptionalism.[5] For example, as Victor Velázquez demonstrates

3 *Renaissance Posthumanism*, ed. by Campana and Maisano, p. 3.

4 See Campana and Maisano, eds., *Renaissance Humanism,* pp. 3–7, for a summary of a few of the ways in which 'Renaissance humanism' has been used as a critical proxy for the ills of anthropocentrism by theorists. Kenneth Gouwens's chapter in this volume engages Carey Wolfe's (mis)characterizations in particular: 'What Posthumanism Isn't: On Humanism and Human Exceptionalism in the Renaissance', in *ibid.,* pp. 37–63.

5 Or, as Phillip John Usher puts it in his important corrective to ahistorical evocations of humanism, the *homo* of early modern humanism is not the *anthropos* of the Anthropocene. See 'Untranslating the Anthropocene', pp. 56–77.

beautifully in his reading of Du Bellay's 'rewilded' Rome, ruins lay bare the temporality of human activity and its relation to—not its distance from—nature. Early modern French humanists were already theorizing natureculture; or rather, they did not have to theorize it because the two terms had not yet been opposed (Sara Miglietti and Oumelbanine Zhiri both engage this Latourian separation directly in their chapters). Early modern French *écologie* emerges in this book again and again as a relational impulse, an ethos, that embeds human bodies and minds always already within vibrant and sympathetic (or antipathetic, as Jennifer Oliver reminds us) networks, rather than separating them from a purified nature beyond human culture.

Early modern naturecultures are about connection, the porosity of boundaries: the Great Chain of Being may establish a hierarchy of creation, but it has *links*, not gaps. This does not, however, mean that Renaissance humanism proposed a celebratory, beautiful discourse of respect for the nonhuman. We should not think magically over distance and imagine that humanist ecologies were somehow closer to a Golden Age of natural harmony. The chapters in the first section on 'Dark(ish) Ecologies', in particular Jennifer Oliver's, provide a bracing corrective to any tendency we might have to think uncritically with Foucauldian resemblance: images of meshed connection can be brutal and violent just as easily as pastoral and tranquil. Similarly, Pauline Goul shows the importance of the Cynic tradition to (dark) ecology, arguing for the complexities of the sardonic laugh that is both joyful and appalled and inviting us to harness our sense of environmental dread to Diogenic thought.

This book's authors show that is it not anachronistic to wonder about Renaissance *écologies*—quite the contrary. Renaissance thought is everywhere situated within the *oikos*; thoughts and texts themselves emerge as ecological actants. Even a proverb, Kat Addis shows, can be a Mortonian hyperobject, connecting the discrete and local with the general and massively distributed. For Stephanie Shifflet, God's contemplation of His manifold creation in Du Bartas is ecological in its 'litany of awe' and the way in which every created thing is rendered beautiful, strange, but also intimate, like Morton's strange stranger. The ecological turn in criticism is not about imposing upon texts meanings that aren't there, as do the readers upbraided by Rabelais in his Prologue. Rather, it is about surfacing relations that have always been there, between text and places, objects, animals, plants, elements. Our readings of Renaissance texts have become so overdetermined by textuality and intertextuality, such an echo chamber of references, that we might have forgotten, quite simply, and as Phillip John Usher reminds us in his beautiful chapter, to see a rose. Rather than always reading for allegory, what if we allow ourselves an encounter with the spectral materiality of Ronsard's rose? *Carpe florem.* Maybe the full effect of the poem is only accessible, Usher concludes, if the reader brushes up against the vestigial rose-ness of the flower, the better to feel the melancholy of decay. If one goal of criticism is to resurface the affective power of writing—and I do hope this is still a goal we share—then reading ecologically can produce the *frisson* of brushing up against something real.

Thus the rather pompous question I asked at the MLA has produced, in the form of the present volume, an essential complication of the question itself. I set up an opposition between ecocriticism and the French Renaissance which immediately begged for nuance, and which might simply have shored up some of the premises of the 'theory wars' in our field so helpfully described and questioned by Hassan Melehy in his chapter. As Melehy collapses the distance between textuality and matter in Montaigne, he makes a broader case for communication between past and present which 'involves a humbling, a disposition of setting aside triumphalist attitudes toward the past'. He, and Usher and Goul in the Introduction, question the resistance to 'theory' from some scholars of early modernity, pointing out that pure historicism—the notion that an authentic understanding of the past is recoverable without taking into account our own subject position—is as much of a theory as anything else. It is true that my wondering how French Renaissance culture might interrogate ecocritical theory (rather than the other way round) constitutes a reversal of polarities, but it is perhaps still too much within the 'application model'. In other words, to posit a critical encounter in binary terms is to frame the parameters of our work as always moving between what Campana and Maisano call the 'two poles of almost irresistible attraction':[6] either we show how the Renaissance anticipates modernity, or we show how modernity is not modern. Rejecting the pole of our own modernity—ignoring the fact that we dwell in an academic world that is having urgent conversations about ecology, and yes, 'race and gender' too—is not an intellectual third space that transcends the polarity, but simply another way of confirming it through the force of rejection.

Ecocritism and the French Renaissance, then, are not poles at either end of a temporal or conceptual spectrum, but are each part of a continuum, a long dialogue in which we humans are emmeshed with the vibrant matter of the world we inhabit. Thus the impulse of this book is perhaps—and in

6 Campana and Maisano, eds., *Renaissance Humanism*, pp. 7–8.

the best sense—quite conservative, in that we seek to conserve continuity between past and present, between text and world, and to take seriously the primacy of the object of study. I was immediately struck by how all of the volume's authors practice and model close reading as a way to encounter, or simulate (following Deleuze and Montaigne via Melehy) material relations. I would also argue that *conservation* is at stake in early modern French *écologies* and our appreciation thereof: conservation both of nature and of… the humanities. This may be a stretch, but I appeal here to Pierre de Ronsard and to a forest he loved.

When Henri de Bourbon became king of Navarre in 1572, he consulted with the council of Vendôme about how best to reduce his family's debts. The council recommended the sale of the Gâtine forest, a place beloved by Ronsard. The forest was sold, surveyors were hired, and trees started to fall the next year. Ronsard published an elegy in 1584 lamenting the clearcutting of an oak forest. This poem, which does not mention the Gâtine by name, was nevertheless given in 1624 the title 'Elégie contre les bûcherons de la forêt de Gâtine', by which it has mostly been anthologized to this day. The question that I and other readers have asked elsewhere is: what to do with the trees in Ronsard's twenty-fourth elegy?[7] Are we primarily to read them as historical referents to real trees felled in a real forest, the Gâtine, and Ronsard's grief as a proto-environmental protest against the loss of natural beauty or an ecosystem? Or are they mostly allegories for classical poetry, the Muses, and a higher moral universe? It is true that Ronsard quickly modulates to the allegorical, even saying 'these are not trees':

> Escoute, Bucheron, arreste un peu le bras,
> Ce ne sont pas des bois que tu jectes à bas,
> Ne vois-tu pas le sang lequel desgoute à force
> Des Nymphes qui vivoyent dessous la dure escorce?
> Sacrilege meurdrier[.][8]
>
> (Listen, woodcutter, hold your arm right there, | This is not a wood which you cut down, | Don't you see the fast flow of the blood | Of the Nymphs who used to live under the thick bark? | Sacrilegious murderer).

7 Mackenzie, *The Poetry of Place*, Chapter 5.

8 Ronsard, 'Elégie XXIV', published originally in the *Œuvres* of 1584. Cited from Ronsard, *OEuvres complètes*, ed. by Céard, Ménager, and Simonin, vol. 2, pp. 408–409. My translation.

And in the subsequent extended *adieu* to the forest, the space is crowded with pastoral *topoi,* satyrs, Pan, lovelorn shepherds, and Ronsard himself composing lyric poetry:

> Adieu vieille forest, le jouet de Zephyre,
> Où premier j'accorday les langues de ma lyre,
>
> (Goodbye, old forest, Zephyr's plaything, | Where first I tuned the tongues of my lyre).

However, there is no doubt in my mind that Ronsard is also genuinely distressed at the transformation of a real place dear to him. The uptake and success of the spurious title naming the Gâtine forest shows the affective power of mourning actual trees. Readers since 1624 have clearly responded as much to this sense of ecological loss, as to the sweeping sense of loss of an old world of classical values and literature, a world of humanist ideals. As with Usher's rose, it is maybe the loss of real natural objects that moves readers the most. But there is no need for an either/or habit of reading: we can read bifocally, for the trees as well as for humanism and poetry. Perhaps Ronsard's 'Adieu, saincte forêt' has become today's proclamation of the 'death of the Humanities'. Trees and poetry were disappearing together for Ronsard; for us today, both the planet and the humanities are in crisis, and these crises might have something in common. Maybe Ronsard's humanism and today's equally beleaguered humanities share something ecological, a commitment to balance, sustainability, relationality?

It is an odd time to be a humanist. An open-ended Google search for the keywords *humanities in crisis* as I write theses lines (in September 2018) yields over thirty-two million (!) results. It is true that, like the always-rising bourgeoisie, the humanities seem to have always been in crisis; as Blaine Greteman reminds us, Robert Burton was lamenting the banishment of the Muses back in 1621.[9] Are humanists defined in part by a perpetual posture of defensiveness, a feedback loop of notional crisis to which we respond by creating more conditions which we then experience also as crisis? Perhaps we dialectically produce the humanities as a perceived refuge from what Du Bellay, in the sonnet I started with, calls 'the wretched business of acquiring more', and what twenty-first-century humanists might call late capitalism, the neoliberal academy, or STEM-focussed administrators. It is true that the current ringing of our own death knell is, in part, a twenty-first

9 Greteman, 'It's the End of the Humanities as We Know It And I Feel Fine'.

century reformulation of a stance that has always defined us, a sense of being somehow outside, but also an essential counterbalance to, this wretched business of making a living. Of course, humanists are not situated magically outside anything; we—those of us lucky enough to have a job at least—are implicated in the same economic logics as the administrators and business schools we love to scorn.

But it does not feel hyperbolic to talk of a particularly existential threat to the humanities in our current moment. Most of us who have been housed in a humanities, arts, language, or social science department since 2008 in the USA would agree with Benjamin Schmidt that this time, crisis talk is not simply crying wolf.[10] There's not much arguing with the plummeting lines in Schmidt's grim graph of enrolment numbers, and students contemplating potential lifelong debt have every right to expect their degree to fast-track them to a good job (the extent to which STEM or professional degrees actually do that is another question, and the essay defending the transferable or 'soft' skills of humanities curricula has become something of a subgenre of crisis literature).[11] The humanities are at a tipping point, in the USA at least, and while the harm of climate change, also at a tipping point by all expert accounts, is exponentially more than the harm of humanities curricula disappearing, the two are surely linked.

In its scorn for accumulative logics, Du Bellay's 'miserable business of acquiring more', from the sonnet with which I started, is not unlike the contemporary humanist's distrust of business models in academia. Instead, Du Bellay dreams of sustainability, a modest *oikos*, and we dream of universities that teach the humanities because they are good to think with, not because they will 'acquire more' for our institutions. Ronsard's anguished lament at the loss of classical poetry, and of the felled trees, has become our lament at the loss of humanistic study, the departments of French and philosophy and drama and German falling like… trees. Behind each of these Renaissance pleas for humanist values, is an ecological thought, and I think the same might be true today. The humanities and the planet are both in crisis for similar reasons. To lose the humanities is to lose something of the

10 Schmidt, 'The Humanities are in Crisis'.

11 For example, Adams, 'Majoring In The Humanities Does Pay Off, Just Later'. As Schmidt points out in 'The Humanities', when tuition is free and students are guaranteed a job after graduating (i.e. at military academies in the USA), the humanities fare much better. I would add that the perception of a crisis is particular to the USA, and to a lesser extent the United Kingdom and Canada. In France, the question itself is often framed as an American concern. For example Louisa Yousfi argues that the value of the humanities is a particularly American (i.e. US) debate linked to late capitalism, in 'Pourquoi enseigner les humanités?'.

subject position of the early modern *homo*,[12] an interrogation of our relations with—not our detachment or distance from—human and more-than-human others. Humanism and the humanities offer a sense of connection to the planet we inhabit: a connection, as these chapters have shown, that has always included the nonhuman, offering us ways to think and live *with* and *in* our teeming world. And *écologie*, as Pauline Goul enticingly put it to me, becomes a 'not-only human-istic form of empathy and care for and among bodies in crisis, be they human, non-human, or fully abstract'.[13] May we all, urgently, keep caring.

Works Cited

Adams, Susan, 'Majoring In The Humanities Does Pay Off, Just Later', *Forbes*, 22 January.

Campana, Joseph, and Scott Maisano, eds., *Renaissance Posthumanism*, ed. by (New York: Fordham University Press, 2016), p. 3.

Du Bellay, Joachim, *Les Regrets et autres œuvres poétiques*, ed. by M. Screech and J. Jolliffe (Geneva: Droz, 1979).

Gouwens, Kenneth, 'What Posthumanism Isn't: On Humanism and Human Exceptionalism in the Renaissance', in *Renaissance Posthumanism*, ed. by Joseph Campana and Scott Maisano (New York: Fordham University Press, 2016), pp. 37–63.

Greteman, Blaine, 'It's the End of the Humanities as We Know It And I Feel Fine', *The New Republic*, 13 June 2014.

Mackenzie, Louisa, *The Poetry of Place: Lyric, Landscape, and Ideology in Renaissance France* (Toronto: Toronto University Press, 2011).

Nardizzi, Vin, 'Remembering Premodern Environs', in *Object Oriented Environs*, ed. by Jeffrey Jerome Cohen and Julian Yates (punctum books, 2016), pp. 179–183.

Ronsard, Pierre de, *OEuvres complètes*, ed. by Jean Céard, Daniel Ménager, and Michel Simonin (Paris: Gallimard, 1994).

Schmidt, Benjamin, 'The Humanities are in Crisis', *The Atlantic*, 23 Aug 2018.

2014.

Usher, Phillip John, 'Untranslating the Anthropocene', *diacritics* 44 (2016), 56–77.

Yousfi, Louisa, 'Pourquoi enseigner les humanités?', in 'De la pensée en Amérique', *Sciences humaines* 17 (2012), pp. 60–63.

12 Usher, 'Untranslating the Anthropocene', p. 62.

13 E-mail to author, 12 October 2018.

About the Author

Louisa Mackenzie (Associate Professor of French, University of Washington) grew up in Scotland and did her graduate research in Berkeley, California. Her research focus is primarily on early modern French culture, which she reads through various contemporary critical lenses including ecocriticism and, more recently, Animal Studies. Her book *The Poetry of Place: Lyric, Landscape and Ideology in Renaissance France* (University of Toronto Press, 2010) is an interdisciplinary study of how a subjective and affective sense of place was produced by poetry in dialogue with cartography, land use history and other knowledge spheres. She is also the co-editor of *French Thinking With Animals* (Michigan State University Press, 2015). Her presentation on an MLA panel in Vancouver (2015) was central to the concept and shape of the present book.

Index

Page numbers in *italics* refer to images

For Product Safety Concerns and Information please contact our EU
representative GPSR@taylorandfrancis.com
Taylor & Francis Verlag GmbH, Kaufingerstraße 24, 80331 München, Germany

www.ingramcontent.com/pod-product-compliance
Ingram Content Group UK Ltd.
Pitfield, Milton Keynes, MK11 3LW, UK
UKHW020016200926
475702UK00014B/1923

* 9 7 8 1 0 4 1 1 7 8 4 6 0 *